Why Organizational Change Fails

Routledge Studies in Management, Organizations, and Society

This series presents innovative work grounded in new realities, addressing issues crucial to an understanding of the contemporary world. This is the world of organised societies, where boundaries between formal and informal, public and private, local and global organizations have been displaced or have vanished, along with other nineteenth century dichotomies and oppositions. Management, apart from becoming a specialized profession for a growing number of people, is an everyday activity for most members of modern societies.

Similarly, at the level of enquiry, culture and technology, and literature and economics, can no longer be conceived as isolated intellectual fields; conventional canons and established mainstreams are contested. Management, Organization and Society addresses these contemporary dynamics of transformation in a manner that transcends disciplinary boundaries, with books that will appeal to researchers, student and practitioners alike.

Why Organizational Change Fails

Robustness, Tenacity and
Change in Organizations

Leike van Oss and
Jaap van 't Hek

Routledge
Taylor & Francis Group

LONDON AND NEW YORK

First published 2011
by Routledge

Published 2014 by Routledge
711 Third Avenue, New York, NY 10017

Simultaneously published in the UK
by Routledge
2 Park Square, Milton Park, Abingdon, Oxfordshire OX14 4RN

Routledge is an imprint of the Taylor and Francis Group, an informa business

First issued in paperback 2015

Typeset in Chaparral Pro by IBT Global

Library of Congress Cataloging-in-Publication Data
Oss, Leike van, 1964–
    Why organizational change fails : robustness, tenacity and change in organizations / by Leike van Oss and Jaap van 't Hek.
        p. cm. — (Routledge studies in management, organizations, and society ; 13)
    Includes bibliographical references and index.
1. Organizational change.   2. Organizational behavior.   I. Hek, Jaap van 't, 1953–   II. Title.
    HD58.8.O84 2011
    658.4'06—dc22
    2010048742

ISBN 978-0-415-88619-2 (hbk)
ISBN 978-1-138-95991-0 (pbk)
ISBN 978-0-203-81590-8 (ebk)

Contents

Figures

Tables

Preface to the English Edition

This book was originally published in the Netherlands in 2008. Since then, we have noticed that our insights regarding the difficulty of change also struck a chord with colleagues from other parts of the world. This gave us the idea of translating the book[1] and making it accessible for an international audience, and you have the result of this process in front of you now.

Since the publication in Dutch, three years have passed and a lot has happened surrounding our book. Our thinking on the subject has also developed and gained more depth. But it still revolves around the core that we described three years ago: the idea that organizations have a robust center, which serves as a kind of immune system protecting them against planned changes that are too rash and too far beyond the organization's own system.

Paul Valens made the following comments on the book at www.managementsite.nl:

> *The breadth of the questions Van 't Hek and Van Oss touch upon is considerable, both in terms of themes and practical illustrations. But owing to their consistent system of observation and description, this is not a problem at all. As in Geert Mak's Jorwerd[2] you quickly get used to making considerable leaps of a geographical, historical, business-related and theoretical nature. Within the space of 155 pages we spend time with an action group, the Netherlands' Directorate-General for Public Works and Water Management, our own families, the paper manufacturer Van Gelder Papier, typewriters in 1873, the French revolution, Vladimir Putin, a candy factory, a housing corporation, a ministry, a Chamber of Commerce, a major supermarket chain, and a council of Mayor and Aldermen. We also join operational managers on an away day. And we make some notable forays into the literature: Books you may have read in a cursory way are concisely and clearly*

---

1   With thanks to UvA Talen, and especially Sarah Welling for translating the manuscript.
2   Geert Mak is a historian who connects great historical events to the reality of ordinary lives that form part of that larger history. Jorwerd: The Death of the Village in Late C20th Europe (Harvill Press, 2000) is a good example of this method.

reintroduced. The references to other reading materials form a rich and varied complement. You can put the book aside at any point and then pick it up again when you wish, and even start reading at a point that is completely different from where you left off. And sometimes you can just decide to skip a section, leaving it for later . . . What does unchangeability look like? A book. And that is what it is, but it is also a print out of a website, a brain-teaser, and an encyclopedia. As well as a catapult, which launches all your preconceived notions up in the air, time and time again. Sometimes they come back down unchanged. Sometimes they are completely revised. And some preconceptions will have disappeared for good.

We could not have asked for a greater compliment. Now the book has been made available to an international audience we hope it will offer you the same experience as a reader. We are certainly very curious and eager to find out to what extent our insights are typical for the Netherlands and to what extent they are recognizable and useful in other parts of the Western world. That is why we will be very happy to welcome reactions to our book at onveranderbaar@organisatievragen.nl

Jaap van 't Hek
Leike van Oss
Utrecht, Maarssen, 2010

Preface

Change is obligatory; standing still equals going backward. Markets are dynamic, the economy is increasingly subject to globalization, regulation is either being tightened or loosened, the competition is gaining ground, the labor market is shrinking or growing, the new CEO wants to make his or her mark, the shareholders are getting restless, the product range is becoming outdated, or the company is being taken over. We need to move forward, to grow and develop in order to be capable of reacting adequately to continually changing demands. Moreover, in most cases we need to act at double speed.

Outside the field of organizational management we see much evidence of skepticism regarding the potential success of such changes. One editorial in one of the Netherlands' leading newspapers included the following commentary on proposed plans for slimming down the central government under the leadership of top civil servant Roel Bekker: *"We need to ask ourselves whether slimming down government bureaucracy does not usually have the same effect as slimming has on the human body, that of eventually making it expand to an even greater size."*

Of all the changes that are initiated, a large number fail. In his inaugural speech on his acceptance of the endowed chair of Organizational Change Management at the University of Amsterdam, Jaap Boonstra (2000) stated that more than 70% of changes either stagnate prematurely or fail to produce the intended result. The reasons for such failures are generally sought in the change strategy, in characteristics of organizations, or in people's reactions to change. The aim in identifying these issues is to resolve them and implement changes more effectively. We too examine these phenomena in this book; however, instead of regarding them as obstacles in the path of change, as they have often been approached in literature on the subject, we see them as expressions of the unchangeable side of organizations.

It goes without saying that change is something valuable, and it is undeniably true that organizations develop. They change from within or in reaction to a process of change. However, they will only ever change in part. If an organization were to be changeable in all its aspects, it would be like a weather vane without a pole or a wheel without an axis: such an organization would lack any form of character, memory, or value. The unchangeable side of organizations is an essential component.

When we argue that organizations are unchangeable we do not mean to say they are unchanging. The Merriam-Webster Online Dictionary defines "unchanging" as "constant, invariable." It describes an incapacity for change and is thus not applicable to organizations. But we can use "unchanging" to refer to the intentions of an agent who implements change: a manager or a director, possibly assisted by a consultant, intends to change an organization in a certain way or in a certain direction. More often than not, such intentions fail: organizations prove to be unchangeable in a number of aspects.

The Focus of the Book

We wrote this book for people who regularly deal with changes and who are interested in the question of why such changes do not always work out as planned. Change strategies are not the central focus of this work. What is central is the way in which organizations construct a robustness within themselves that is not easily changed, and how that robustness can result in tenacity in a process of planned change.

In order to enable us to adequately describe the relevant manifestations of unchangeability, we have made use of theories that throw light on the way in which an organization constructs and maintains itself, as well as theories about the way in which people contribute to these processes and act within these frameworks. These are mostly social-constructivist theories. They help us to gain insight into the agents who build organizations and their construction processes. This perspective shifts the focus from the change agent, the change process, and the object to be changed to the system and its internal dynamic. This enables us to describe unchangeability as a system characteristic.

Unchangeability in organizations is a subject that is relatively unfamiliar and difficult to approach for change agents. It is for this very reason that we decided to include perspectives from other disciplines and fields of knowledge in this book—fields in which thinking about change is less determined by the a priori ideology of change than in the field of change management (its name says it all). We were fortunate enough to find a philosopher, an ethologist, an ecologist, an archeologist, a linguist, a movement scientist, a biologist, a sociologist, and an architect who were each willing to write a brief essay on change and unchangeability in their particular fields. Most of the chapters in this book open with one of these essays. The aim of these stories is to stretch our views on unchangeability through examples taken from other disciplines.

A Guide to Reading

This book consists of three main parts. In Part I, which is titled "Robustness" and consists of Chapters 1–5, we describe robustness as a characteristic of organizations. This is the capacity organizations have to form and maintain themselves, and to which people contribute through their actions. Robustness is a characteristic that is always present. This is not to say that it is rigid: robustness has both a stable and a dynamic side, and together these create a sturdy balance, which is difficult to change. We describe this balance in Chapter 1 and focus on a number of different aspects of robustness in the chapters that follow it.

In Part II, which is titled "Tenacity" and consists of Chapters 6–9, we describe the phenomenon of tenacity. Tenacity is a manifestation of robustness that arises in reaction to change strategies. It becomes visible when processes of change do not go as planned and the results of a change trajectory differ from what was expected. Interaction between the robustness of organizations, people's actions, and the change strategies applied can give rise to tenacity. We describe the way in which people manage, either consciously or unconsciously, to get change initiatives to spring back, to smother them, or to use them to pursue or safeguard their own interests.

In Part II, titled "Perspectives," consisting of Chapters 10 and 11, we look at why it is that unchangeability is such a little-examined subject and what knowledge about unchangeability can mean for change.

We wrote this book intending it to be read from the beginning to the end; however, feel free to choose the reading path that best suits you. The introductory essays from different fields and the chapters can be read independently. If you are mainly interested in information about unchangeability in organizations, you could choose to skip the introductory essays. If you want to make your way through the book even more quickly, you might find it useful to focus on the more theoretical chapters, "Robustness" and "Tenacity." If what you are looking for is some brief sketches, metaphors, or parallels derived from other disciplines to help you understand unchangeability, then why not browse from one introductory essay to the next. If you are looking for some background information on social constructivism, you can turn to the appendix we included dedicated to this subject.

All the cases included in this book have been borrowed from real-life situations and events. We include them by way of example and have therefore mostly revised them in such a way as to make it impossible to trace them to particular individuals or organizations.

A Word of Thanks

We could not have written this book without the inspiring contributions of our co-authors: Hans Bennis, Piet Boonekamp, Christien Brinkgreve, Frank

Bijdendijk, Saskia van Dockum, Jan van Hooff, Theo Mulder, René Gude, and Louise Vet. They helped us and will hopefully help the reader to gain a better understanding of the phenomenon of unchangeability through their differing perspectives, and for that we offer our heartfelt thanks.

Less visible to readers, but nonetheless of great value to us, was the feedback provided by people who read earlier drafts of this book, including Adrie van den Berge, Margot van Bergen, Frank Bakema, Yola Claasen, Marjo Dubbeldam, Jos Heijke, Erica Koch, Elisa Koehof, Miel Otto, and Martijn Sillevis. Without them, this book might well have been finished earlier, but it would definitely not have been the same finished product.

Our thanks also go out to our clients and their employees who allowed us the free range of their organizations for many years, enabling us to see how some things change and others do not. This has allowed us to build up a wealth of experience and a large share of the insights we have included in this book. We are exceptionally thankful to those people who have listened to our theories on unchangeability over the past two years and have provided us with their critical responses: they have helped us to sharpen our views.

Despite the fact that a great many people have contributed to the thoughts that have shaped this book, ultimately we of course remain entirely responsible for its contents, with the exception of the contributions by our co-authors, who have kindly provided their own texts.

March 2008, Leike van Oss and Jaap van 't Hek

Part I

Robustness

In Part I we describe the phenomenon of robustness: an organization's capacity to shape and protect their individual character. It is the characteristic feature that sets an organization apart. In the first chapter we set out a theory of robustness, and in the following three chapters we go on to describe the way robustness is formed through social, cognitive, and political aspects of sensemaking, respectively.

1 Robustness

During the time that we were writing this book, we took a walk through a landscape that one of us knew from his past, but where he had not been for more than thirty years. To his delight and surprise, nothing had changed. This was obviously not entirely the case, as many of the trees that stand there now had not been there thirty years ago. Why, then, did he perceive the landscape as "unchanged"? The answer to this question lies in the fact that the fundamental pattern of the landscape has not changed. The alternation between forest and meadow is the same; the mixed forest of oak, beech, birch, larch, and pine is the same; the tranquility is the same; the sandy footpaths are the same. Preserving the landscape in this state requires maintaining the fundamental pattern of paths and occasionally replacing a bench. Things do change, therefore, but such change is usually in the interest of what is already there. Nature lends a hand as well. During these thirty years, old trees have died but new ones have appeared. The effect is a sense of continuity.

Obviously, many more changes occur in nature than the occasional death of one tree and the growth of another. Each season brings its own changes. We were there in the fall; six months later, in the spring, it would have looked entirely different. This cycle—this perpetually recurring change—also provides a sense of continuity.

In organizations, we can observe a similar continuity, which appears to be founded on a balance between adaptation and stability. If you look back over an organization's history you will see that many things have changed, while many others have not. In their book, *Spelen met Betekenis* (*Playing with Meaning*), Ruud Voigt and Willem van Spijker (2003) discuss change in government organizations and explore various approaches and the factors that contribute to the success or failure of these approaches. The authors describe how change and preservation go hand in hand:

> We work with organizations that are marked by innovation and continuity and that, like ourselves, are struggling with these forces. In our stories, we hope to show the great extent to which government organizations are changing. At the same time, bureaucratic organizations are characterized by stubborn continuity. In some cases, this could serve as evidence that nothing fundamental has changed in these organizations. We are of the opinion, however, that bureaucracies, with all their certainties and fixed rituals, offer considerable space for innovation. There can be no innovation without structure. The certainty and security of several elements of bureaucracy allows people in government organizations to accept radical change. (p.249)

It is not only in historical perspective that we can clearly see how much has remained the same despite the many changes that have taken place. The same balance can be observed in the present.

For example, consider the secretarial pool in a large organization, in which a group of secretaries are working for a number of managers and policy workers, each with a unique division of tasks and a fixed way of working. Within these patterns rests the capacity to respond to changes that occur each year. For example, the secretaries know that they will receive an enormous mountain of work right before most of their "customers" leave for a vacation, but this will be followed by several fairly calm weeks. This time can be used for archiving. They also know that the hectic Christmas-card period will begin in November and that everything will be relatively calm again around Christmas. They are aware of events that recur each year, and they adapt their working methods accordingly.

Incidental adaptations take place as well. For example, new managers often bring different questions and demands. These questions and demands are incorporated into the existing way of working. None of these events or external demands changes the core of the work pattern, but they do lead to adaptation.

We use the term *robustness* to refer to the balance between stability and adaptation in the interest of such stability. Because of robustness, the fundamental pattern of the organization remains the same. Robustness is therefore the capacity of an organization to retain its core characteristics under changing conditions. For this to be possible, robustness must have two sides: existing patterns and routines, on the one hand, and the capacity to create and preserve them, on the other. In some situations, organizations need to be able to adapt to external influences in order to guarantee stability.

Robustness is a capacity that remains invisible unless we focus on it. Robustness is what the organization normally is, in the same way that the landscape normally exists. The robustness of the landscape is noticeable only when something has changed, or when one returns to it after many years, only to find that very little appears to have changed at all. The same is true of robustness in organizations: it is only when our attention is drawn to it that we notice it.

1.1 Change and Unchangeability as System Characteristics

Robustness is a capacity. Describing robustness requires theories that provide insight into how organizations build and maintain themselves. To this end, we draw upon insights from social constructivism, as this perspective provides insight into the ways in which people jointly create organizations and shape them into cohesive, stable entities. These notions can also be used to describe how it is possible for change and unchangeability to exist simultaneously in organizations.

From the social-constructivist perspective, people create their own realities through interaction with others. In the context of organizations, Karl Weick

(1979) refers to these processes as "organizing." According to Weick, organizing is the core activity with which organizations are built and maintained. Organizing consists of the processes of sensemaking, which people use to interpret information from their surroundings and transform their interpretations into a created reality. This reality subsequently provides direction for behavior, which reinforces the created reality.

Sensemaking consists of three phases (Weick, 1979): enactment, selection, and retention. Sensemaking begins with the moment at which people notice something in their surroundings that is unfamiliar to them. According to Weick (1995), this involves three types of disturbances: unexpected occurrences, occurrences that are expected but that do not take place, and processes that are interrupted. In each of these cases, attention is drawn to something that is unfamiliar and to which a meaning must be assigned. The drawing of this attention marks the enactment phase of sensemaking, the phase in which information is actively identified and selected. Weick refers to this process as "bracketing." In the selection phase, information that has been placed "in brackets" is positioned alongside schemas that exist in people's minds, in order to arrive at an explanation for the phenomenon. The explanation that is found is registered in the retention phase. The retention phase subsequently provides direction for what was identified in the enactment phase and what served as comparative material in the selection phase. Not every disturbance is noticed. Only that which is registered as a construct in the retention phase and which actually allows space for identifying the disturbance is noticed. Sensemaking is therefore a self-reinforcing and self-confirming social process.

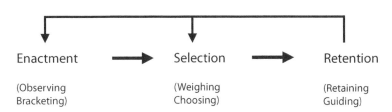

Figure 1.1   Sensemaking (adapted from Weick, 1979).

Sensemaking entails both the capacity to change and the capacity to preserve. Active enactment and selection allows people to use sensemaking to adapt their realities and to change their behavior accordingly. What is registered in the retention phase ensures the selective identification and interpretation of information. Sensemaking therefore has a closed character, which reinforces individual views of reality and serves to strengthen rather than to change.

Sensemaking is thus both a source of change and a source of preservation and continuity.

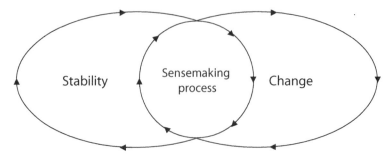

Figure 1.2 Stability and change.

As shown in Figure 1.2, both capacities—the capacity to bring about change and the capacity to build stability—are always present. In this chapter and throughout this book, we focus on the latter capacity: the capacity to build stability and to preserve.

1.2 The Formation of Robustness

Robustness is the capacity of an organization to remain the same in terms of its core characteristics under changing conditions. Every organization develops this robust capacity, which is applied to build, strengthen, and preserve the organization. As the process becomes more ingrained, robustness increasingly becomes a characteristic of the organization.

It is seldom possible to observe how robustness develops within an organization from the very beginning. In order to understand robustness well, however, it is helpful to know the process through which it is shaped. Further insight into processes of sensemaking can help provide a better overview. In the rest of this chapter, we will therefore delve deeper into this subject and conclude by describing a practical example of the emergence of an organization and its robustness.

1.2.1 *Sensemaking*
People are constantly looking for ways of making sense of the phenomena that occur around them. Doing this makes the world clear and understandable to them. Meaning is an indispensible element in this process. Without meaning, the world is incomprehensible to people and therefore uncertain and unmanageable.

> Imagine that your colleagues occasionally break off their conversations and retreat to a conference room when you enter the room. They never invite you to join them, and they are secretive about what they are discussing. You would obviously like to know what they are discussing and whether it involves you. When your questions are met with evasive reactions, you become insecure. Left

WHY ORGANIZATIONAL CHANGE FAILS

in the dark, you start coming up with your own explanations for their behavior toward you in isolation, and your performance is likely to suffer as a result.

Sensemaking is less an individual activity than it is a collective form of information processing. Together, people are continually passing through the three phases of enactment, selection, and retention in order to make the world understandable. It is this interaction that leads to the convergence of personal images. In this way, a common, shared reality emerges—a created world that is not objective but constructed socially by people. What is created, in its turn, determines the direction of people's behavior. Since people have created a common meaning together, they also know how they should act within this meaning, and this makes it easier for them to be effective within it. People are thus both the object and the subject of their own construct: they play a role in a story that they write themselves.

Imagine that you are travelling with a group of people that you do not know in an inhospitable area, off the beaten tourist track. You camp and eat together, and you need each other in order to travel around. In the initial stages of the trip, the group searches for ways of camping, eating, and traveling that are suited to the area and for the best way in which to work together. Each form of behavior that contributes to this effort is retained. The behaviors that prove unsuccessful are discarded. In this way, the group discovers which behaviors are appropriate, and they interpret the surroundings and the contributions by and relationships among members. In this way, routines emerge that the group members can feel confident about and which are effective in the surroundings in which they find themselves. In the process of acting, they find out which one of them is best at making fires, at cooking food, and at setting up camp. They also know whom to avoid in the mornings or how to deal with one group member who tends to get dispirited. They divide the different tasks among themselves and avoid encroaching on each other's terrain. They also accept leadership and guidance from each other on the basis of this division of tasks. As subjects, and in the course of their travels, these people have given shape and meaning to their cooperation. At the same time, they are also the objects of this form of cooperation. It is important for them to continue to display the behavior that has proven effective. What has been created thus determines the direction of their own behavior.

The frequent interaction also leads to the emergence of emotional and problem-solving experiences that are recognized as mutual. As observed by De Moor (1995), *"We come to realize that we somehow belong with each other. A common awareness of mutuality and single-mindedness emerges."* People become attached to what has been created, and they adhere to the corresponding rules of the game.

1.2.2 Constructs
The result of sensemaking processes consists of constructed images of reality: constructs. That which is registered is highly diverse, as is the form of the

constructs. Weick (1995) conducted an inventory of concepts appearing in the literature that he considered to fall under the category of construct (with regard to meaning). The list includes ideologies, third-order controls, paradigms, theories of action, and stories.

- Ideologies are *"shared, relatively coherently interrelated sets of emotionally charged beliefs, values, and norms that bind some people together and help them to make sense of the world"* (Beyer in Weick, 1995, p.111). They describe the values and norms that provide direction for behavior within organizations.
- Third-order controls are constructs that, according to Weick, include the "taken-for-granted assumptions" of the organization (Schein). These assumptions act as unconscious control mechanisms for behavior. Weick also refers to them as "professional blind spots." They include convictions that are assumed to be true.
- Paradigms consist of broad intellectual visions—worldviews—into which fundamental cognitive schemas and values are closely woven. Paradigms are the convictions that provide direction for professional behavior.
- Theories of action (Argyris) are sets of rules that individuals use to explain their own behavior, as well as the behavior of others. Theories of action are cognitive organizational structures that direct the behavior of people within organizations.
- A tradition is a conviction from the past that is passed along to subsequent generations. Within a tradition there are convictions, experiences, and behaviors that are anchored in the past. Traditions help to transmit these convictions, experiences, and behaviors to new generations.
- Stories are present-day guidelines for people in organizations. Becker (1988) describes the emergence of stories as follows: *". . . that things don't just happen, but rather occur in a series of steps, which we social scientists are inclined to call 'processes' but which could just as well be called 'stories'. A well-constructed story can satisfy us as an explanation of an event."* (p. 31)

All of the types of constructs that Weick mentions contain stories about the world around us, as well as instructions for acting within this world. A construct provides direction for how we observe the world around us and how we think about it. It provides guidelines for how we should react to it and how we should act accordingly.

> Several years ago, the Foundation for Idealistic Advertising in the Netherlands (SIRE) produced a commercial that was aimed at making people aware of the way in which they apply certain images almost automatically. The beginning of the film shows a man with a Middle Eastern appearance sitting on his knees bowing forward and sitting back up a number of times in succession. The immediate association this image evoked in viewers was of a Muslim at prayer. The producers of the commercial clearly anticipated this reaction. After a while, the camera pans out to show that the man is paving a street.

One advantage of constructs is that they tell us how to act. Recognizing the situation removes the necessity of figuring out what is happening, thereby conserving information-processing capacity. People like to use constructs as a frame of reference for their actions. Constructs form a pattern or "script" that can be drawn upon to provide information about the situation.

In some cases, such constructs are so tight or so self-evident that their content transcends discussion. For example, in many organizations, "growth" is accepted as an important criterion for success. Growth is good and worthy of being pursued. Why this is true for a specific situation, however, is not always clear.

> In an address at a meeting of an interim-management company, the director expressed pride that the company had grown by 30% in the previous year. The keynote speaker for the evening, the Dutch philosopher Bas Haring, questioned whether such growth was actually a good thing. A number of interesting discussions arose about this question during the closing drinks. Although many participants maintained the standpoint that growth was good, few actual arguments could be found to support this standpoint.

The process through which constructs become implicit and the way in which they come to be beyond discussion is the same process that Pauka and Zunderdorp (1988) describe in their book *De Banaan wordt Bespreekbaar* (Opening the Banana to Discussion). In their story, a group of monkeys know that they will be soaked with a hose every time they dare to reach for the bananas that hang from the top of their cage—or at least the original inhabitants of the cage know this. New inhabitants learn primarily that they must not climb up to the bananas. Although the monkeys do not know why this is not allowed, the message is quite clear: "Don't touch the bananas!" Even when the situation is such that none of the monkeys in the cage have ever experienced the direct link between water and picking bananas, they still do not climb up to the bananas. Their behavior has become self-evident. None of the monkeys has ever encountered the actual motivation that lies behind the behavioral rule (avoiding a soaking), but the prohibition against touching the bananas is nonetheless upheld.

A close connection exists between constructs and the sensemaking process. One cannot exist without the other. The process of receiving information from the environment, placing it within the frame of reference provided by the previously retained information, and preserving or adapting one's image of reality transpires continuously. Constructs are always present as well. We develop them in the course of our lives, as we are confronted with new situations. They are imparted to us as we are being raised by the society in which we grow up, through television, through films, or through our own experiences. All of these constructs guide which information is identified, as well as the way in which it is interpreted.

Constructs constitute the stable side of robustness, while sensemaking processes constitute the dynamic side.

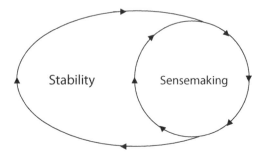

Figure 1.3   Robustness.

1.3   Operational Closedness: The Unchangeable Aspect of Robustness

Robustness is a stable, unchangeable characteristic, as previously formed constructs come to form a filter for what we observe in our environment. We see what we are meant to see according to the construct through which we are present. Robustness makes an organization an operationally closed system. The fact that it is operationally closed does not mean that no information enters the system. It simply means that information is selectively identified or interpreted. Information that is consistent with the organization's own identity and interpretations is most readily identified. Information that contradicts the organization's identity and interpretations is not seen, or it is interpreted in such a way as to make it consistent with these interpretations.

> Consider the following example: a documentary showed a Dutch member of parliament visiting a young IT company. The small space was populated by informally dressed and coiffed young wizards surrounded by cans of soda, leftover pizza, and computer screens. The politician's critical questions regarding occupational safety regulations and work and rest periods were met with incomprehension from all sides. The employees did not feel at all disadvantaged by the lack of such regulations. They talked about their interesting projects and their willingness to work hard to complete them. They enjoyed their work and had obviously created a pleasant working environment together. On leaving the building, the politician stated that the working conditions were a real issue. The image he held was not changed by the factual information he had observed (e.g., people were happy; they were experiencing no ill effects resulting from their lack of proper protocol regarding work and rest periods). Consequently, his behavior in parliament revealed no visible evidence that he had adjusted his convictions in light of this information.

In this example, the politician's identity and inner world consist of his convictions about organizations and working conditions. These convictions guide the way in which he interprets the information. Any information that is liable to contradict his convictions is not included in the evaluation of the situation at the IT company, or is weighed in such a way as to make it consistent with the politician's own convictions (and identity). The company's employees on their part interpret the politician's questions in relation to their own world. In this context, the message implied by the politician's questions is out of touch with the young IT workers' experiences.

In a robust system, information and impulses from the environment are not considered and used objectively; they are interpreted, and the system acts according to this interpreted information. This also applies to impulses for change. After having been interpreted, such impulses do not always lead to change.

> For example, consider an organization in which management considers "customer-focused operations" an important issue for change, while the employees on the shop floor are convinced that they have already been operating in a customer-focused way for years. The process of change that is set in motion is likely to be met with considerable resistance, due to the existence of a "closed" notion of customer focus, which serves as a filter for explaining the process of change. In situations of change, operational closedness thus functions as a filter between the change agent's intentions, on the one hand, and what an organization actually does with these intentions, on the other.

1.4 Three Aspects of Robustness

Weick (1979) describes sensemaking as a cognitive and social process. Sensemaking processes are cognitive, in that sensemaking is a learning process and knowledge emerges within this learning process regarding the environment and the way in which organizations and individuals relate to this knowledge. Sensemaking processes are social, in that people create meaning together through social interaction.

Hosking and Morley (1991) add a third aspect: the political aspect. Sensemaking processes are political, as interaction involves not only cognition and interaction, but also the establishment of relationships between parties.

Since people use sensemaking processes to create and preserve their image of the world, the constructs that emerge from these processes also comprise three aspects:

- The cognitive aspect is expressed in memory. Memory is the historical and layered web of knowledge people have about their environment and the organization.

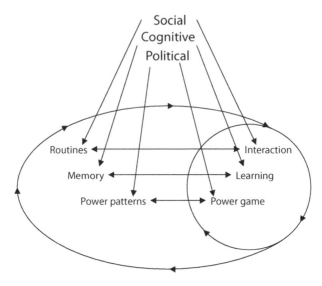

Figure 1.4   Aspects of robustness.

- The social aspect is expressed in routines. Routines are patterns of behavior that guide people's actions.
- The political aspect is expressed in a pattern of power. Patterns of power are the "solidified" relationships in organizations, the patterns of mutual relationships.

The following table contains a schematic representation of these notions:

Table 1.1   Aspects of Robustness

Aspects / Facets	Social (chapter 2)	Cognitive (chapter 3)	Political (chapter 4)
Dynamic (process)	Interaction: the interaction that takes place between people within the frame-work of routines.	Learning: the continuous process of testing alterna-tive patterns of action and retain-ing successful ones.	The power game: the game that is played according to the rules that are produced by the power pattern.
Stable (construct)	Routines: habits and alternative courses of action.	Memory: the historical web of knowledge about the organization and its environ-ment.	Pattern of power: stable pattern of power relations and interactions in an organization.

The preceding description seems to suggest that the facets and aspects of robustness can be clearly distinguished from each other, diagnosed analytically, and followed separately. In practice, however, constructs and sensemaking processes are closely intertwined, and cognitive, political, and social aspects all play their roles in organizations simultaneously. They alternate in importance in the various events, and they are evaluated differently by the various parties involved. In short, the phenomena that we have so carefully unraveled and attempted to describe are actually intertwined and are not always clearly distinguishable. Nonetheless, people are able to find their way in this complex world on a daily basis. To do so, they order the world in a number of ways.

The first ordering principle is the sensemaking process itself. This process helps people to create meaning out of disorganized information—meaning that is clear and makes sense to them. Selecting from all available information and evaluating this selection in terms of our own actions makes the environment less ambiguous.

A second ordering principle resides in the fact that sensemaking is always a social process. Meaning emerges through interaction, and this process of creating meaning also connects people to each other (De Moor, 1995). What results is a group of people who relate to the created reality, to each other, and to the patterns of behavior that emerge. Homan (2006) refers to this as a local meaning island: an informal group with which one identifies and within which the members are continually constructing their own reality together. On this island, people feel connected, and this connection serves to strengthen constructs. In organizations, meaning islands can be departments, groups who have lunch together, or even people who meet each other in the smoking area. They can reflect either the formal or informal face of an organization.

We believe that, if they endure, meaning islands ultimately become "islands of meaning": constructs of meaning that manifest themselves as "stories" within the organization and to which people may or may not relate. Whereas meaning islands still consist of people, "islands of meaning" exist independently. People can relate to a number of these stories and be part of more than one island of meaning.

The final ordering principle is what Homan refers to as the organizational landscape. The organizational landscape is a subjectively ordered map that provides direction for the behavior of people. It is a composite of people, islands of meaning, and stories that are more or less interrelated as "loose" entities with regard to the three aspects of robustness.

An organization consists of not one but many islands of meaning, which may be located closer to or farther away from each other, or which may overlap each other to varying degrees. As people in organizations share meaning with each other and continually encounter each other in interaction, the islands of meaning of which they are part also become associated with each other. According to Homan, there are two types of links between islands of meaning that determine the closeness and the nature of the relationship. Closeness is determined by the number of links between islands of meaning, as well as the intensity of these links. Closeness is also determined by the nature of the relationship between islands of meaning. The greater the number and intensity of the links between different meanings, the closer they are brought in proximity to each other. A greater divergence of meaning emerges, thereby increasing the closeness between the islands. Interaction between people is the foundation for this closeness. Two entities within an organization that are brought into intensive contact with each other through their work share a great deal of meaning with each other and influence each other's view of reality.

The nature of the relationship is also important. It tells us something about the way in which the meanings of the various islands relate to each other. And this, in turn, is determined by power relations. The meaning of the most powerful island is thus more "true," or is proclaimed more loudly within the kingdom of islands than that of the less powerful island. The meaning that is produced by the "directorate" island has a greater impact on the "mailroom" island than the other way around.

People know the organizational landscape and the constructs within it. Sensemaking allows them to bring some kind of order to this landscape for themselves. That is why people experience robustness as a very familiar, well-understood, and therefore manageable whole. It is not an objective fact but a subjective form of knowing that has arisen through interaction and that is transmitted to new people entering the organization in an implicit way.

> For example, at the start of your first day in a new job, you have a rough idea of what is expected of you. You will probably start at nine o'clock; you know you will have a lunch break, and you have a general idea of the kind of work you will be doing. You may already be familiar with the sector and therefore with some of the prevailing mores. You will be less familiar with the customs and habits, the history and the relationships within this specific company. During the first few months, your colleagues will instruct you by telling you stories about how the organization works and providing valuable advice about how to behave within it. These stories and this feedback on your behavior teach you what the organizational landscape and the islands of meaning look like.

1.5   Where Does Robustness Reside?

Robustness is a "substance-less" characteristic in the sense that it is not restricted to a specific component or facet of the organization, but should rather be seen as a capacity that has implications for all aspects of the organization, although it is more visible in some facets than it is in others. Robustness is closely intertwined with the behavior of people, and it is therefore more visible in those components of the organization in which the people's behavior plays a central role.

For example, formal organizational structures are less closely associated with human behavior than is organizational culture, and teamwork is more dependent upon human behavior than it is on computerization systems. In this context, components that are less dependent upon human behavior obviously do require human behavior in order to function optimally. In essence, however, human behavior affects some facets of the organization more and others less.

Organizational culture is closely tied up with human behavior. With its norms, values, and behavioral practices, organizational culture may even appear to coincide with robustness, but it is not the same. Robustness is an organizational capacity, while organizational culture is an organizational cha-racteristic—a characteristic in which robustness is most clearly expressed.

The fact that robustness is closely interwoven with the behavior of people provides insight into the extent of changeability or unchangeability of various facets of the organization. Where human behavior plays a central role, the capa-city of robustness is more crystallized than it is in components in which human behavior plays a less central role. An organizational chart can be changed rela-tively easily, and a computerization system can be implemented quickly. Mat-ters become more difficult when people must work within such new structures or with such new systems. As an unchangeable characteristic, robustness plays a particularly important role in matters relating to human behavior and relati-ons; it has less to do with the more "technical" aspects of the organization.

1.6 Organizational Development from Scratch: A Grassroots Case Study

The emergence of robustness in organizations is difficult to explain using examples of existing organizations. In existing organizations, the capacity for robustness has already generated crystallized and robust characteristics. It is possible to describe the robustness of existing organizations as a phenomenon; the description of the way in which it actually came to be quickly becomes spe-culative in the context of existing organizations. Fortunately, this book is able to draw upon an example of a grassroots organization that recently developed out of nothing and with which one of the authors was closely involved: the Plat-form against the A6–A9, an organization that arose out of a collective desire on the part of civilians and environmental organizations to put a stop to plans to construct a new highway. We will use the formation of this organization to il-lustrate how the capacity for robustness results in stability. We will spread this over four chapters. In this chapter we begin with the process through which the organization developed into an existing, functioning system. In the fol-lowing chapter, we elaborate on the cognitive, social and political aspects of the organization.

> The Platform against the A6–A9 (hereafter referred to as the Platform) came into being when *Rijkswaterstaat*, the Netherlands Directorate-General for

Public Works and Water Management (hereafter referred to as RWS), presented its plan to improve mobility between Schiphol airport and the city of Almere. One of the most obvious alternatives was a new highway that would connect the highways numbered A6 and A9. The creation of this highway would have a major impact on the landscape and living environment surrounding Amsterdam. A number of groups rose in opposition to this proposal, and the Platform came into being as an umbrella organization for these groups, all of which were affected by the RWS initiative in their own ways.[1] All of these groups thus had differing concrete interests, accents and views with regard to the RWS plans. These interests included protection of the environment, of the landscape, of the living environment, and of agriculture in the area. Nonetheless, they also started out with one common interest in which their separate interests converged: a concern for the quality of the area in a broader sense. These groups united themselves in the Platform, thereby taking the first step toward a new organization.

In terms of sensemaking, the situation was one of a number of local but relatively unrelated islands of meaning, which were brought closer together by actions on the part of RWS, both in terms of behavior and sensemaking. The RWS resolution affected relevant areas in all of these local communities. The impulse from RWS caused the islands of meaning to discover a greater correspondence in meaning than they would previously have anticipated before RWS became a "common enemy." The local islands of meaning thus approached each other while maintaining their own local identities at the same time.

In the interactions within the individual towns or associations, the local meanings remained intact, while a shared, overarching identity was developed in the members' interactions with each other in the Platform.

Two factors played a role: the image that the outside world had of the Platform and the way in which it was possible to transcend local meanings to arrive at collective meaning.

The image that the members of the Platform had of how they were perceived was that they ran the risk of being portrayed as Not In My Back Yard types, or NIMBYs: rich, highly educated people from the Gooi region with extensive networks—well-to-do people who were unlikely to be considered deserving victims in the eyes of the public. This awareness affected the Platform's behavior, and it had an influence on the norms, values, and convictions that guided their actions. They felt it was important to ensure that they did not provide a confirmation of a stereotypical view that people could conceivably have of them. This motivated the members to immerse themselves in the issue so that

---

1   As the Platform's "opponent," RWS is obviously a simplification of the complex reality. RWS was merely a player in a political field in which the minister, political parties, social organizations, the press and other entities were all of influence. From the perspective of the Platform members, however, RWS never succeeded in behaving as "neutrally" as it had presented itself, and it was perceived as an opponent throughout the entire game.

they could show that they were not merely opposing something in their own backyards, but had a constructive attitude and had alternative suggestions. This affected their media strategy as well. For example, members who lived in large houses avoided being photographed in front of their homes.

A second effect was that this immersion generated the insight that the mobility question was too complicated to resolve merely by laying down more asphalt. In fact, the conviction grew that more asphalt would only lead to more traffic congestion. The Platform decided that their efforts should therefore extend beyond opposition to one particular freeway; it became their mission to show that an extra freeway was not the solution, regardless of the location. This conviction led the members to seek contact with residents of Amsterdam Zuid-Oost, who formed another group of potential victims of the RWS plans. If the Platform had been content merely to fight a battle to keep the highway out of their area and had succeeded in this, the residents of Amsterdam Zuid-Oost could have lost out instead. The fact that the Platform opposed the plan to build a highway regardless of the location made these potential adversaries into allies in pursuit of the same cause.

In the process of acting and building upon a common interpretation, a Platform emerged with a specific identity, transcending the local islands of meaning. This subsequently ensured that all of the groups could relate to the Platform identity. The Platform acquired the subtitle "For innovation in mobility."

Once the Platform had a clear identity, a clear interpretation of the problem and a clear position in the field, this served to establish and reinforce the image that they presented to the outside world, as well as their own behavior. The image eventually gained depth both within and outside the group, in part because of the necessity of answering critical questions posed by journalists, of writing position papers, and of "working" the relevant members of parliament. All of these actions forced the members of the Platform to engage in internal discussions before presenting their case to any external audiences. These internal discussions developed, reinforced, and deepened the internal organization.

1.7 Summary

Robustness is the capacity of an organization to retain its fundamental pattern and core characteristics under changing conditions. It is a capacity that is expressed in all aspects of the organization. Once an organization has existed for some time and robustness has become manifest in all of these aspects, it becomes an inseparable characteristic of the organization.

As a capacity that generates unchangeable aspects of an organization, robustness does not replace change but is rather an equal counterpart to it. Both change and robustness are always present. Sensemaking processes connect the two capacities; although sensemaking processes can lead to adaptation and change, they can also be focused on the retention and reinforcement of that which is already there.

As an organizational characteristic, robustness is so self-evident that it is not immediately visible. We seldom see how robustness takes shape in practice. We usually encounter existing organizations in which robustness has already taken shape. Only in situations in which new organizations emerge is it possible to track this process, of the formation of robustness, as well. In this chapter, we have attempted to construct (or reconstruct) this process.

2

Routines:
The Social Aspect of
Robustness

Habits and routines are important elements of robustness. Once established, routines are hard to change. This calls to mind the way in which men and women, despite the desire for equality, are repeatedly confronted with patterns that point toward inequality. This is an issue that not only affects individual relations, but has consequences for the way in which society is organized. This example prompted us to ask Christien Brinkgreve, who has been writing on this subject for years, to give us her considered opinion on what she sees as changeable and unchangeable in the relation between the sexes.

On the Tenacity of Sexual Differences

Prof. Christien Brinkgreve

Professor of Social Sciences at the Utrecht University

Once upon a time, we thought and hoped that men and women were equal, in a sense that went beyond equal rights, equal opportunities, and equal recognition (rewards). We also saw equality in terms of qualities and capabilities. The differences that clearly existed and had been emphasized and cultivated for centuries were explained away by referring to the influence of culture, power relations, and the way in which people were raised—the influence of "socialization in gender-specific roles," as it was called in the 1970s. Paying attention to differences was considered suspect, as it could easily serve to legitimize social inequality. Women were capable of just as much as men were, and they were just as valuable. After all, because physical power was no longer an important source of power in our post-agricultural and post-industrial society, this source of male dominance had also lost its force and rationale.

All sorts of measures were undertaken in the pursuit of gender equality: scenarios were written detailing the equal division of work and care duties; blueprints were drawn up listing target quotas for a more equal division of functions and positions at work; and advertising campaigns were launched to prepare people mentally to desire such equality and give it a try.

Despite these aspirations, however, several differences have proved stubbornly persistent. For example, although many have embraced the ideal of an equal division of labor and care, such a division has proved difficult to attain in practice. There continue to be good reasons for men to keep their full-time jobs and for women to take a step back: men usually earn more, and women actually do tend to be more adept at managing children.

Attaining equal positions at work is another such aim. Formal barriers that discriminated against women have been lifted, and employing women at the highest levels is seen as modern and progressive. Nonetheless, women are able to break through the glass ceiling only at great effort and, having broken through, they often tumble back down again relatively quickly. "They didn't quite fit in with the team, after all," or "She wasn't professional enough"—the power to define situations has remained a male prerogative. After having been brought in because emotions were deemed important, women have demonstrated the wrong kind of emotions, indicating weakness instead of the strength that was actually deemed necessary within those companies.

Attempts were made to tackle the roots of inequality by raising children in a gender-neutral way from the very beginning, or even bending them slightly in the opposite direction. Even though boys were given dolls to play with and girls were given cars and swords, the difference failed to stick. The behaviors and preferences of children continued to follow the old patterns.

Part of the explanation can be sought in evolutionary psychology: there are biological differences between men and women (in build, hormonal systems, and brain structure), and these differences have taken a more specific shape in the course of the history of humanity, in reaction to the physical and social environments in which people have found themselves. Changes in power relations and culture can undoubtedly lead to changes in behavior and feelings, and they can either magnify or reduce differences between the sexes. For example, women's liberation has resulted in more women working outside the home and wanting to have careers, regardless of whether they have children. Nonetheless, differences in preference remain. In general, men are more oriented toward work and career, and women are more oriented toward home and family. Although this does not mean things cannot be the other way around, and although it does not belie the fact that there are many variations on this pattern, a tenacious difference tends to remain between the sexes with regard to these tendencies, especially when children are involved. Even when good arrangements are in place to help people continue to work full-time (for example, good childcare facilities and ample provisions for parental leave, as in Scandinavian countries), women still retain a stronger preference than men do for participating in the care of their children.

Several decades later, differences between the sexes are proving more stubborn than expected. These differences are also important for organizations. The sexes demonstrate different reflexes when confronted with the threat of danger: women look for safety in good relationships with others, while men tend to fight. Men are more sensitive to ego damage and the loss of rank and status, while women are more sensitive to the loss of ties and their significance to others, as well as the recognition they receive. Men strike back, while women retreat to safer ground. Men feel safe when they have power and control, while women derive a sense of security from good relationships with others. Women have a greater capacity for empathy, while men are better at dealing with conflicts at work and are less inclined to take such incidents personally. Men are more problem-oriented and rigid; women ponder things for longer, have a better eye for the context, and are more flexible.

For organizations, these differences are important, but they are also tenacious, stubborn, and not likely to disappear quickly in response to changes in culture (either societal or organizational). Culture obviously does make a difference, however: the space available for different capacities and qualities and the recognition that is accorded to them determine whether people will survive, thrive, or languish within a given working environment. It is therefore better to recognize the differences than to force people into the same mold of uniformity. Recent history has taught us that it is possible to change inequality between the sexes. The position of women has changed radically in recent decades. Nonetheless, social patterns are not as easy to change as we might like them to be.

2 Routines: The Social Aspect of Robustness

In the previous chapter, we listed three aspects of robustness: cognitive, political, and social. This chapter addresses the social aspect. Like the others, this aspect has both a stable and a dynamic side. The dynamic side consists of interactions between people, which form the basic building blocks of the sensemaking process. The stable side consists of the routines that people create together through interaction.

Table 2.1   Aspects of Robustness: The Social Aspect

Aspects / Facets	Social	Cognitive	Political
Dynamic	Interaction	Learning	Power game
Stable	Routines	Memory	Power pattern

2.1   Interaction: The Dynamic Side

2.1.1   *It Takes Two to Tango*

Karl Weick (1979) calls interaction between people the cornerstone of sensemaking. Without interaction, individuals would be left to process information in isolation. It is interaction that transforms sensemaking into a thoroughly social process, and it is interaction that ensures that meaning forms a coherent web instead of a collection of separate meanings. It is because people find themselves in unorganized surroundings that contain a diversity of information and stimuli that they feel a need to "organize" these elements into coherent wholes. In order to do this, argues Weick, people need each other.

According to Weick, the act of organizing can take place only if two or more people collectively try to make sense of events occurring in their surroundings. Interaction takes place through a process that Weick calls "interlocked behavior": behavior that connects people to each other. Person B reacts to something that person A has said or done. Person A subsequently reacts to person B's reaction. In this way, people demonstrate that they care about other people's reactions to what they do or say.

> Imagine that you are angry with a colleague who has failed to send a document as promised. You storm into your colleague's office and launch into a tirade. Your colleague reacts to your anger by apologizing and giving an indication of why it was not possible to follow the agreed-upon procedure. Up to this point, the scenario consists of action and reaction. No adapted, shared meaning has yet emerged, although there is insight into each other's motivations. After you react by indicating that you understand your colleague's

position, accept the apology, and ask for advance warning should things not go according to plan in the future, you know that your relationship has been restored. More importantly, you know that each of you has taken a step forward in making sense of your reality, the way in which you relate to each other, and the way in which you want to work together. It is only at this point that we can identify a circular effect in your communication with each other. A new rule has been agreed upon: provide due warning when you cannot finish something you promised.

Robustness cannot be developed without this social aspect. Meaning is agreed upon through interaction, and this ultimately results in communality.

2.1.2 Making Rules for Sensemaking

Interaction allows people to organize, and thereby eliminate, uncertainty. In some cases, this process is aided by existing rules and routines; in others, people must discover these procedures through interaction with each other. The more complicated and unfamiliar the environment, the greater is the lack of clarity, the volume of uninterpreted information, and the amount of interaction needed to clarify that environment. In the process of interaction, meaning is attributed to the environment, and rules and routines for dealing with that interpreted environment are constructed at the same time.

Groups that have been together longer and environments that are more familiar involve a greater accumulation of rules, thereby demanding less interaction in order to create new meaning. In this case, interaction serves primarily to reaffirm existing realities.

In a close-knit department in which people have been working together for a long time, employees implicitly know "how things work around here," and they know each other through and through. When certain situations arise, people are familiar with the routines that are called for. This knowledge is confirmed by every action carried out by anyone within the department and by every reaction to these actions from within the environment. The self-evident and implicit nature of the interaction within the department does not become visible as such until a new colleague comes in who is unfamiliar with these routines. This new colleague constantly asks questions that appear obvious to the others. Conversely, when we find ourselves in the position of newcomer, we must connect with others in order to uncover the rules and routines that they have already developed for situations with which we are not yet familiar.

2.1.3 Actions and Beliefs

Karl Weick (1995) argues that sensemaking can be initiated by action and by beliefs, for which he uses the terms "action-driven sensemaking" and "belief-driven sensemaking." Belief-driven processes are sensemaking processes that stem from the beliefs that people hold with regard to the world, as these beliefs are stored in people's minds in the form of ideologies, values, paradigms, and similar notions. These constructs and the beliefs that inform them guide the things people notice and the way in which they interpret events. As Weick states, "To believe is to notice selectively" (p. 133).

One example of a belief-driven sensemaking process is provided by the case of the politician in the previous chapter: he saw what he wanted to see based on his views and convictions. Action-driven sensemaking stems from actions: in the course of their actions, people are confronted with the way in which they experience the world around them and the way in which the world reacts to their experiences. This process can lead people to adjust their meaning.

Imagine a situation in which you are responsible for giving a presentation to an important group of people, and your presentation misfires. There are long silences, the discussion you had planned to begin falls flat, and the participants fail to relate to what you are telling them. After such an incident, you search for an explanation for the failure of the presentation. This is an example of action-driven sensemaking.

Whether meaning stems from actions or beliefs has consequences for the kind of connections that people will forge with one another. People are selective. It is not possible to engage everyone in interaction about everything in all situations. People select potential interaction partners according to the resources that these potential partners have to offer (Weick, 1979; Termeer, 2001) or because they define reality in the same way. This process leads to the creation of the islands of meaning that we described in the previous chapter.

2.2 Routines

Interaction gives rise to routines—patterns of behavior that guide action. Routines consist of the habits, methods, rules, and procedures that arise over time. These rules become so self-evident to people that they are no longer aware of the fact they are applying them.

> Imagine you are visiting your parents along with all your siblings (assuming you have siblings). Before long, you all fall back into the patterns of your youth. This may even go so far that you all sit in the same places at the table, your brother decides which games to play or which channel to watch, and your parents react to your "squabbles" in the same way that they always have. These are all familiar routines that evolved as you grew up. They are contextual, but they are also your own. Although they organize behavior within your own family situation, you also carry the routines that you have acquired within your family into other situations, and these routines help shape your images of reality. For example, you might react to the behavior of an authoritarian boss in the same way that you used to react to your authoritarian father. If you were the youngest child in your family, you might assume a comparable position in other groups. You might even find that you feel most at home in positions and organizations that mirror your family situation in some way.

Routines can also be made explicit. Working methods, procedures, protocols, and job descriptions are all examples of routines that have been made explicit. People act in accordance with these procedures, at least for as long as they are workable. Routines are usually workable when they are the result of people making sense of their reality by interacting with each other. In many cases, routines are less workable when a given working procedure has been created from the perspective of an island of meaning (for example, the management island) that is relatively unrelated to the meaning of the island that carries out the work. In such situations, tension arises between the formal description and informal routines and working procedures turn out to be "paper realities" instead of explicit routines. The interaction between paper intentions and their actual realization thus usually leads to a hybrid between the paper design and actual practice.

> The staff members of a research institute were used to developing knowledge in dialogue with each other and their surroundings. This was a largely organic process, in which the researchers divided up areas of attention among themselves and maintained client contacts on that basis. The institute's management determined that client relations should be formalized, and they decided to introduce account management and a corresponding information system. This was a well-designed routine originating from another island of meaning, but one that only bore a limited connection to the existing routine that people had developed communally. As a result, it was difficult to get the new role of account manager and the new management information system off the ground in practice. Because this working method was imposed on them, the researchers ultimately could not avoid dealing with it. They took some external aspects from the management's wishes and transplanted them to their own island of meaning. In this way, they did manage to realize a form of account management and the new information system, but within their existing routines and interaction. The researchers now decide on a case-by-case basis who will be the account manager and how this team member will maintain contact with the group with regard to current acquisitions.

Routines from one part of an organization can influence routines in another part. If we view the management style within an organization as a routine (whether implicit or explicit) developed by the organization with regard to management, this routine is likely to share a remarkable number of characteristics with the primary processes managed by the organization. For example, we observed how the interviews regarding personal development plans took the form of negotiations within a labor union and how career-development talks within a police organization involved superior officers issuing developmental "orders."

The influence of the primary process can also be observed in policy. For example, legal security and equality before the law are key concepts in the staff policy of the civil service, and we found that a child-protection agency considered the reports of staff performance interviews to be confidential and thus destroyed them after a period of one year. In some cases, the characteristics of clients also become entangled with the ways in which colleagues interact with each other. In one addiction agency, staff members were quick to use blackmail tactics on each other and to confront each other with *faits accomplis*, in much the same way that their clients treated them.

These examples show how routines that are developed in the primary process and are thus considered almost self-evident (and that are therefore less well considered) can be transferred to an organization's management routines and policies, or to the ways in which people work together internally. The self-evident character of routines and the fact that they help to reduce ambiguity are the primary reasons that they are almost automatically transposed onto other aspects of organizing within organizations.

2.3   Configurations

In the previous chapter we discussed islands of meaning. These arise from social configurations people form with each other. A social-cognitive configuration comprises a (temporary) group of people who share a particular view of reality: *"What characterizes a configuration is the presence of intensive interactions between certain individuals who share the same definitions of reality for a large part"* (Termeer, 2001, p. 371). People can belong to multiple configurations. Moreover, configurations are temporary; they take shape and eventually fall apart again. The difference between islands of meaning and configurations seems to be that the latter have a more temporary character than the former. Nonetheless, both function as "structures" within organizations. While routines provide direction and guide people's actions, configurations and islands of meaning provide structure.

2.4   Organizational Culture

Routines and patterns of action guide people's behavior in relation to their own environments. In organizations, we can view organizational culture as an encompassing web of routines and patterns. Proceeding from a social-constructivist description of culture, it is possible to distinguish both the stable and the dynamic components of the social aspect of robustness within organizational culture.

De Moor (1995) provides the following social-constructivist description of culture: *"an organization-as-social-reality, which is seen as an inter-subjective reality which is continuously constructed for and by its members as they communicate and interact"* (p. 106). De Moor makes a distinction between the organizational ideology and the organizational climate. This difference relates to action-driven and belief-driven sensemaking processes.

The organizational ideology is the coherent system of ideas that are characteristic of—and binding upon—a social unit or organization. De Moor sometimes refers to this as a "social frame of reference," a coherent system of assumptions, values, norms, views, ideas, and preferences that help to determine what the evolved community deems desirable. Belief-driven sensemaking processes are important for the development and maintenance of organizational ideology.

According to De Moor, ideology guides people's actions. In the process of organizing, they develop artifacts and practices by choosing actions collectively. He views artifacts and practices as expressions of the dominant ideology. According to Weick (1979, 1995), these practices also arise in the opposite order, as in an ideology that emerges out of action-driven sensemaking processes.

Figure 2.1   Organizational culture (De Moor, 1995).

In this context, culture can be understood as a broader web that encompasses separate patterns of action and routines.

2.5   Individuality Within Robustness

In interaction, people are driven not only by routines and patterns of action that are constructed through interaction, but also by their own individual systems of convictions, norms, values, and capabilities. People are systems in themselves, with their individual sensemaking processes, their individual histories (which have given rise to meaning), and their own alternatives for action, convictions, and feelings, which act as important motivations for behavior.

According to Gregory Bateson (1984), people function on six distinct neurological levels:

- environment: the stimuli to which individuals react
- behavior: the concrete conduct of individuals
- capability: skills and aptitudes
- belief: the things that individuals hold to be true, their central values, and convictions

- identity: individuals' descriptions of how they see themselves
- spirituality: individuals' personal perceptions of the meaning of life.

These levels shape the ways in which people define themselves in relation to their surroundings and determine the motivation of their actions. On all six levels, people make assumptions and learn. All six levels are involved in the individual sensemaking process; together, they form an individual's coherent system of meaning. This influences the kind of collective meaning that arises, as well as the way in which individuals relate to this collective meaning.

De Moor (2005) argues that individual meanings converge in collective meaning. Through intensive interaction, individuals' norms, values, and convictions, along with their personal repertoires of actions, become part of a collective repertoire of actions. This is not to say, however, that individual meaning is swallowed up in the collective whole. People retain their own, individual meaning systems, which form the basis for determining the ways in which they relate to the collective.

This process generates more than simply a collective unity of constructs, routines, and patterns of action; it also generates a dynamic between individual and collective systems of meaning. It is because of this interaction that what appears as a robust system on the outside is actually an intricate complex of individual actions and collective patterns.

It is the individuality of people that makes them capable of intentional action, with regard to both the robust context and interaction within that robust context. It transforms people into active players in robustness. This gives rise to a diversity of images and realities.

2.5.1 Emotional Unchangeability

Emotions are one of the characteristics that make people unique actors within systems. Emotions make an individual into more than simply an anonymous part of a greater whole.

According to De Moor (1995) individual sensemaking consists of a complicated process of interplay between thinking, feeling, and acting. When people create their own realities, these realities subsequently have consequences for the feelings that they arouse. In their turn, these feelings affect the actions that people connect to their realities.

Emotions play a particularly decisive role in connecting people to particular realities, even if these are old realities that rooted in the past, which people recall with nostalgia. For example, in old industrial companies, employees sometimes long for a return to the days in which the company was of major importance to the region and a source of economic vitality.

The way that a situation in the present is experienced can be negatively informed by a past reality. Although the situation no longer exists, the negative emotion can still have a direct influence, resulting in a pattern of action that continually allows people to hold on to that negative emotion.

In one factory, the warehouse staff felt the sales staff had a rather vain and condescending attitude. This negative emotion led the warehouse people to treat the sales staff with such a degree of aloofness that the sales staff inevitably reacted by keeping their distance. This was then quickly interpreted as arrogance and condescension, reinforcing the existing negative emotion. The situation between the sales and warehouse staff was never really sorted out. The emotions were too precious to release.

Robustness has a degree of stability in itself, and the individual emotions that are linked to it strengthen its unchangeability.

2.6 Attachment and Freedom

People become attached to routines and patterns because of their stabilizing effect. They also provide freedom. Because we know what the world looks like, we know how to navigate through it. If you have spent your entire career working in government organizations, you will be familiar with the rules of the game and know your way around. If you then move to a commercial organization, you will encounter different rules. You will need to discover these rules. Initially, you will experience this as a constraint until your familiarity with the rules increases to the extent that you can move freely through them. It is these qualities of attachment and freedom that give people in organizations a stake in keeping these organizations the way they are.

2.7 Organizational Development from Scratch: Developing Routines

Because the Platform against the construction of the A6–A9 developed from scratch, it had to make sense of its own position and the world surrounding it through interaction between the people and organizations that formed the Platform. In the initial situation, the Platform was a collection of individuals who had not yet become a close-knit group with patterns of action and routines. Like the local groups, the core group turned out to consist of a highly diverse collection of people. The differences were great, in terms of both political preference and the interests that people were advocating (landscape, nature, historic value, preventing clutter in the living environment, ecology). What everybody had in common was a shared interest in stopping the highway; this common interest provided a sufficient basis on which to build an organization through interaction.

By reacting to the image that others had of them, the group developed the identity described in the previous chapter. This identity was the axis on which interaction rotated, and it formed a foundation for the development of patterns of action and routines. This identity was rooted in the fact that the

Platform was a subsidiary occupation for all the participants and that it was fed by other islands of meaning, including local authorities, a number of legislators, and the media attention that the issue had generated.

In the initial situation, the Platform consisted of a group of individuals who had sought each other out in reaction to current events. At some point, an e-mail correspondence arose between one individual and the other core members regarding a particular issue. The active e-mail correspondence that resulted produced a conclusion. This working method was such a success that the number of meetings was reduced, and e-mail became the most important internal medium for communication. It often allowed the Platform to make sense of a new development in the course of a single day.

As this method of organizing continued and proved successful, it became an automatic routine for reacting to the environment. Because communications with the environment were intensive and frequent, people were able to organize themselves before any formal impulse was launched from outside. For example, the group was provided with a draft of a report by a journalist, long before RWS released the report. By the time that a formal impulse arose on the part of political parties, RWS, or the media, the members of the Platform had already prepared themselves to respond.

In this example, the organization developed through action without a large number of formal functions or a rigid demarcation of roles. The group had a chair who also served as the Platform's secretary. For the rest, a rough division of roles had been agreed upon with regard to acting as a spokesperson and commenting on the massive amount of information produced by RWS. These roles could be performed by more than one person. Each time there was an impulse from outside, an organization of convenience was formed: those who had the necessary time and the expertise grouped themselves around events in the outside world.

The following example serves to illustrate the group's organized routines and the ability to react to situations in a flexible way. During the preparations for a large-scale demonstration in the Naardermeer area, one group member had the idea to set up a small Ferris wheel to provide visitors with a good view of the stunning natural surroundings. The owner of the amusement ride drove up to the site with the Ferris wheel folded up on a trailer. He stopped short before a small bridge, not wanting to risk driving his heavy truck across it. One of the members of the Platform was there at the site and tried, unsuccessfully, to persuade the man to continue. The driver insisted that he would proceed only on the condition that the Platform would accept liability for any resulting damages. As the member of the Platform just happened to be a construction manager in his daily life, he called the structural engineer with whom he regularly worked out of a meeting. The structural engineer asked all manner of factual questions about steel profiles and thicknesses of the bridge construction, ultimately concluding that the driver had been right and the bridge would not withstand the weight. Because this was a highly unsatisfactory answer under the circumstances, a support structure using steel planking was conceived on the spot and through consultation by mobile phone. The owner passed across the bridge without mishap. It was the unexpected and unplanned availability of knowledge combined with a strong motivation and willingness to take risks that helped the group to surmount this obstacle.

It is interesting to note that, months later, the Minister of Transport and Public Works formulated a plan to explain to journalists assembled at that same site how, in the way she envisioned it, the proposed highway would hardly affect the area at all owing to plans for a tunnel. The journalists were put on to buses in order to visit the site. The buses never arrived at their destination, however, as they had to stop at similar bridge along their route.

2.8 Summary

The social aspect of robustness lays the tracks along which the organization moves forward. It guides day-to-day action, methods, rules, and procedures that have developed in the course of interaction and that lead to self-evident and automatic habits in the organization.

Routines and patterns of action provide people with a sense of security and clarity regarding the ways in which they should act. People like to follow routines and patterns that have predictable results, but they are not unresisting followers of routines or prisoners of patterns of action. People relate to the organizational routines according to their own routines, norms, values, and meaning. Individual emotions and feelings play an important role in the interaction between individuality and collective patterns of action.

Routines and patterns of action contribute to the unchangeability of organizations through their stable character and ability to steer people's behavior. It is through routines and patterns of action that people turn behavior into habits that are difficult to change. The effect of a track that forces a cart in a particular direction becomes deeper and more effective with every cart that passes through.

3

The Learned Organization: The Cognitive Aspect of Robustness

Combined with our capacity for learning and unlearning things, history and memory play an important role in the unchangeability of organizations. This is analogous to the relationship between how we make particular motions and the role that our brains play in this process. Theo Mulder, an expert on human movement, agreed to enlighten us concerning all of the processes that are involved in a seemingly simple action (e.g., raising a cup to your mouth). We also asked him to describe what is and what is not changeable within this process.

Changeable Unchangeability: On Human Motor Control in a Constantly Changing Environment

Prof. Theo Mulder

Director of Research at the Royal Netherlands Academy of Arts and Sciences (KNAW) and Professor of Movement Sciences at the University of Groningen

I can walk forward and backward; I can jump, dance, run, shuffle, and produce all sorts of silly walks. I can pick up a cup with the right hand, with the left hand, while the arm is positioned in all sort of angles; I can even pick it up by using my feet as the main effector organs. I can speak loudly or softly; I can make myself understood, even with a cigar in my mouth. This last example would require me to reprogram around seventy muscles, in addition to adjusting my breathing, all within a few milliseconds. I can do this, you can do it; in fact, anybody can do it.

Evidently, we seem to have no problem with producing an almost infinite stream of movements in order to reach goals in our environment. In fact, motor behavior can be seen as problem solving. We are forced to find solutions for the problems which appear in the continuously changing environment. These solutions, however, are never static but always tailored to the actual requirements.

However, if the environmental constraints are never the same, the solutions can never be the same either. This is an important point since it indicates that motor control cannot be the result of a rigid, hierarchically-organized system generating efferent commands to individual muscles and joints on the basis of motor programs stored in some neural warehouse. Control is, for a large part, non-hierarchical, self-organizing, and driven by multisensory input. Furthermore, the organism never functions *in vacuo*, disconnected from its history and without any knowledge. On the contrary, almost all actions are influenced by knowledge and experience. We have learned how to handle a cup, to ride a bicycle, to write, to play the violin, to dance. Even the simplest actions such as opening a door are influenced by learning. For example, we know when to push and when to pull on the basis of knowledge derived from experience.

Hence, there are no stored motor programs guiding our movements in a top-down, detailed, and muscle-specific manner. Nevertheless, this form of control was considered as a likely explanation for quite a long period. The term *motor program* was popular in the 1960s and 1970s. This term implied a system in which detailed neural programs for controlling movements were stored in the brain. This approach floundered because of several unsolvable problems, particularly with regard to aspects of "novelty" and "storage." The *novelty problem* refers to the fact that, if there were detailed programs for every single movement, we would not be capable of executing any movement that we had never made before, as no specific program would exist for that movement. This point is relevant to the study of organizations, as it suggests that the more hierarchical an organization is, the less flexible it will be. Strict hierarchy is not compatible with flexibility, whether in nature or any other context. The *storage problem* refers to the fact that, if detailed (perhaps even muscle-specific) programs were to exist, such a program would have to exist for every single movement that we have ever made. This would create considerable neural storage issues. Although we cannot say with certainty that storage would pose an insurmountable problem, such a model does appear far from efficient, in theory at least.

We can conclude that motor control does not occur according to a strictly hierarchical principle in which the head office (the brain) holds the information required for the "behavior" of separate muscles. Control is exerted in a far more global manner. Although the concept of the motor program has not yet been consigned to the rubbish heap of scientific thought, currently the term is not used much.

Information is a central concept in contemporary theories that argue that everything in the body is attuned to the reception of information, from within the body itself, as well as from the environment. Our bodies are densely packed with receptor systems that play an important role in movement (eyes, ears, skin, sensors in muscles and joints, balance sensors, speed detectors, and receptors for touch and pressure). The composition of this orchestra of sensory information does, however, change from one moment to the next, according to the context, the task at hand, and the state of the system. The control system must therefore react to continually changing configurations and combinations of information. The important point here is that central control systems need not be aware of these combinations in detail. A considerable share of this information is regulated locally, within the body's peripheral nervous system.

We can therefore conclude that the motor system is a natural system in which details are filled in according to global objectives and by relatively autonomous regulation units that are capable of reacting to the available information in a highly flexible way. For example, a reflex is not an unvarying automatic response to a fixed stimulus (e.g., tapping the hamstring with a hammer) but rather a response that is highly dependent on a large number of factors. The results of hamstring-reflex measurements performed on a person at rest differ from those of the same test performed on a subject who has been given a mental arithmetic task to perform.

The point that I wish to emphasize is that movement can be described only in part as a motor process. Movement actually involves permanent interaction between perceptual, cognitive, and motor processes. This interaction between processes (the influence of which can vary in strength) is dependent on the state of the system and the novelty or complexity of the task at hand.

Flexibility is crucial for an organism's survival. The neural control system needs to have the capacity to enable changes in the output according to changes in the input. It must be able to change existing routines and construct new ones on a continual basis. This conclusion re-emphasizes the system's dependence on the availability of information. If no information is available from the environment or the body, the system cannot function, learn, or change.

The process to which we are actually referring in this regard is self-organization. In the simplest terms: when neural networks in the brain receive information, certain connections between neurons are strengthened and others are not. When information is reiterated, it further reinforces the corresponding connections. In this way, the input that is received provides a foundation for the formation of networks that play a role in the control of certain behaviors. The execution of these behaviors (movements) reinforces the networks. While networks are strengthened by the intensification and reiteration of information, they are weakened by decreases in information.

This principle of growth and shrinkage based on input, variation, and repetition actually provides a rough sketch of the structural foundation of learning and change, keeping in mind that many aspects of learning proceed automatically without any form of supervision.

3 The Learned Organization: The Cognitive Aspect of Robustness

This chapter is concerned with the cognitive aspect of robustness. By cognition, we mean the capacity to process information, as well as the entirety of knowledge, insight, and skills that proceed from this capacity. In addition to the social and political aspects, cognition is a third aspect of robustness.

Table 3.1   Aspects of Robustness: The Cognitive Aspect

Aspect Facet	Social	Cognitive	Political
Dynamic	Interaction	Learning	Power game
Stable	Routines	Memory	Power pattern

3.1   The Cognitive Aspect of Robustness

Imagine that you are learning to drive. Learning is the process through which we acquire skills through practice—through trial and error, by making mistakes, and by doing things right and repeating them. The more you drive and repeat the actions involved, the more experienced you will become, until what you have learned becomes so habitual that it is difficult to recount all of the actions involved in driving a car. All of these actions have become automatic, and your knowledge of them has become part of what we call "memory." In organizations as well, memory contributes to stability, clarity, and familiarity.

The cognitive aspect of robustness includes learning end learnedness: the capacity to deepen the knowledge that is already held as well as that of learning new things. To illustrate how this is a part of the unchangeability of organizations, take the fact that it is easier to learn from new experiences than it is to unlearn that which has been learned in the past. Consider such motions as swinging a golf club or serving a tennis ball. Once learned, these movements are difficult to unlearn. Imagine deliberately trying to forget how to ride a bicycle or to unlearn the "seven times" table. At the same time, each action relating to that knowledge strengthens and deepens that skill or the capacity to retain that which has been learned.

The same is true of organizations: although organizations can learn new things, every action relating to something that has been learned in the past strengthens and deepens that skill or knowledge on the part of the group or organization. This produces familiar images of reality and behavioral patterns that are reinforced with each repeated activity, becoming increasingly habitual.

3.2 Learning: The Dynamic Aspect

Learning is the process through which people construct knowledge based on information they encounter in their environments. It may proceed through the transfer of knowledge: for example, people can learn actively, by following a course or reading a book. Another form of learning involves practicing something for as long as it takes to master it, as when a basketball player practices a particular shot endlessly until it can be reproduced at the exact same angle every time. In organizations and other social settings, however, knowledge is created not only by individuals but also, more importantly, through the interaction of individuals with each other. Studies on organizational learning have demonstrated the importance of shared social contexts, shared meaning, and interaction to learning, especially with regard to things that cannot be easily learned from books or through practice (Lam, 2004). This kind of learning develops through and is anchored in the actions of people. When an unfamiliar situation arises in an organization for the first time, someone will respond by conceiving of or improvising a behavioral response. If it fails, the behavior will not be repeated; if it is a success, it probably will be repeated. This is how people learn; they acquire and create knowledge about the world around them by interacting in their own environments.

3.2.1 Skill and Competence

In part, learning relates to necessary actions and skills. If asked to provide an exact explanation of why he does the things he does, a tailor is likely to omit a number of important considerations, as he may be aware of the routine he follows, but not of the precise considerations that underlie these routines. Experienced drivers are able to drive and engage in all kinds of other activities at the same time. If asked to teach a beginner how to drive, however, an experienced driver will probably not be capable of giving a systematic account of all the actions that are actually involved in driving.

One model that is often used in this context (and for which we have been unable to determine the source) describes how the learning of skills proceeds through the following stages:

1. Unconscious incompetence: in the initial stage, people are not aware that they have not mastered a particular skill. "Ignorance is bliss" is a fitting motto for this stage.
2. Conscious incompetence: as they learn, people become aware that they have not mastered a particular skill. In this stage, people become conscious of their deficiency, although they lack the capacity to compensate for it. As a result, everything tends to go wrong, as people attempt to do things differently but are not yet able to do so.
3. Conscious competence: in the third stage, people learn the relevant skill and gain competence through practice. Their competence is still conscious at this stage, meaning that they still need to be conscious of their actions. The new behavior requires considerable concentration.

4. Unconscious competence: in the fourth phase, the skill has been practiced so much that it can be carried out without thinking. It has now become "second nature."

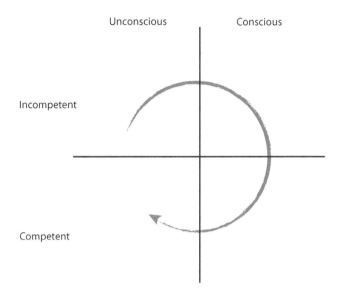

Figure 3.1   Phases in skill learning.

In addition to individual learning processes, these phases can also be applied to constructivist learning in groups. When a situation arises within an organization, people make sense of it through action, search for suitable solutions, and create suitable behaviors in response. Behavior that proves unsuccessful is abandoned, while successful behavior is repeated. This results in conscious competence. The more often behavior is repeated, the more it becomes a habit that can be performed without conscious thought. At this point, members of an organization become unconsciously competent with regard to that situation. Habits can become so ingrained that the original choices underlying the behavior are lost. The routine itself has become self-evident.

3.2.2 *Knowledge*

Learning results in different kinds of knowledge. Lam (2004) proposes a two-dimensional typology of knowledge. The first dimension distinguishes between individual and collective knowledge, while the other dimension distinguishes between implicit (tacit) and explicit knowledge.

Individual knowledge can be applied to questions that confront individuals. When individuals leave an organization, they take their individual knowledge with them, and they can apply that knowledge elsewhere. Collective knowledge is intertwined with the way in which knowledge develops and spreads within a community or communities. This knowledge is accumulated through interaction, and it is stored in the form of the procedures, routines, habits, and standards that can be applied. This knowledge is the "property" of the community that uses it.

Explicit knowledge can be conveyed to others through formal and systematic language. Returning to the example of driving lessons, this learning process involves explicit knowledge about traffic rules, about how cars operate, and about how to apply the traffic rules when driving. This type of information can be conveyed by an instructor and reproduced by a student.

Implicit knowledge is knowledge that people acquire with regard to acting in a specific context, and it is difficult to convey separately from that context. For example, the behavior of a tailor in Turkey or India is likely to differ from the behavior of a tailor in the Netherlands—not with regard to operating the sewing machine, but in terms of responding to the wishes of clients and the local conventions of the trade. It is easy to see how the implicit knowledge of a driver in Turkey or India might differ from the implicit knowledge of a driver in the Netherlands. Implicit knowledge consists of the rules that are learned through interaction with the surroundings and through which skills are developed further. It forms the mental framework within which explicit knowledge is meant to be understood. It provides a backdrop for explicit knowledge (Nonaka and Takeuchi, 1995).

Lam (2004) uses the following schema to present a typology of knowledge, based on the two dimensions described previously.

Table 3.2   Types of Knowledge (Lam, 2004)

Ontological dimension / Epistemological dimension	*Individual*	*Collective*
Explicit	Embrained knowledge	Encoded knowledge
Implicit	Embodied knowledge	Embedded knowledge

- Embrained knowledge is acquired through formal education and training or through such activities as reading books. This type of knowledge is stored in people's brains and is easily transferrable.
- Embodied knowledge is contextual, practical, and action oriented. This knowledge is acquired primarily through experience and in interaction with more experienced colleagues. This type of knowledge is particularly suited to problem solving.
- Encoded knowledge is disseminated through procedures, protocols, working instructions, or other formal systems. This type of knowledge typically involves selections and simplifications.

- Embedded knowledge is communal knowledge that is established in diverse routines and standards. Shared meaning and constructs make it possible to work together effectively. Embedded knowledge is relationship-specific, contextual, and widely disseminated. This type of knowledge is a prerequisite for complex interaction.

3.3 Memory: The Stable Aspect

The mnemonic aspect of robustness is an organization's contextual, implicit, and unconsciously competent knowledge. This aspect encompasses both encoded and embedded knowledge. Encoded knowledge is explicit and therefore objectively knowable. The situation is far more complicated for embedded knowledge, which is implicit. Embedded knowledge is embodied in habits and routines, and it contains the kind of obvious "truths" that people tend to take for granted, as they are implicit and self-evident. Memory evolves over time. It is located and anchored in many different places within an organization. It is not like a central computer that can be accessed from any workstation.

3.3.1 Memory as a Guide to Learning

As we have stated previously, sensemaking processes are continual. This applies to the cognitive aspect of learning as well, given that people are continually faced with new information from the environment. In other words, learning is a never-ending process. Memory does not impede learning. The world changes continually and it continually raises new questions. Memory serves to guide the selection of the information that is observed and how it is interpreted. Just as individuals are always prone to selective perception, social systems of meaning (e.g., organizations) are subject to a similar phenomenon. The more firmly memory is anchored within an organization, the more it guides the way in which problems are solved throughout that organization and the choices that are made with regard to new information from the environment. This process supplements the organization's memory, transforming it into an increasingly robust whole.

3.3.2 Memory: Intertwining Explicit and Contextual Knowledge

The unchangeability of memory can be explained in part by the fact that implicit (contextual) and explicit knowledge are intertwined, as seen by the members of an organization. Such self-evident intertwining allows the members of the organization to grasp the explicit information that exists within the organization and to translate it into action that is meaningful to them. New explicit knowledge has yet to be embedded in day-to-day practice. The connection between explicit and contextual knowledge can be forged only through action.

Consider the way in which the best of intentions, to keep to agreements, are only rarely met. In nearly every organization, failing to honor an agreement is one of the greatest sources of irritation, and it heads the list of areas for improvement. If everyone finds it so important to keep agreements, why is the problem of breaking agreements so difficult to address? In most cases, "a deal is a deal" expresses an intention: everyone is capable of promising (with the sincerest of intentions) that they will try harder. Intentions are far removed from the highly intricate reality that individuals face in practice, however, and reality always presents good reasons for why people are unable to honor their agreements. An intention is "explicit" knowledge that is formulated in response to a problem that has been encountered. The intention, however, has yet to be anchored within the organization's implicit knowledge and contextual routines.

3.3.3 Memory as a Complex, Autonomous Characteristic

The more firmly memory is anchored and the more it forms a complex system, the more difficult it is to change, as the web of interdependencies and interests produce a tight-knit and unchangeable whole. In other words, memory is layered within itself, containing all manner of implicit assumptions, forgotten causes, and implicit self-evidences, and it is linked to the political and social aspects of robustness. As a component of robustness, memory forms an ambiguous knot that is almost impossible to untangle.

Theo Mulder's contribution illustrates the great complexity of individual memory with regard to seemingly simple activities (e.g., lifting a cup). The complexity of social systems (e.g., organizations) encompasses the complexity of all of the complex individuals that are included within them, in addition to all of their interactions, learning processes, and memories. If there is no clearly identifiable center that controls the movements of individuals, as Mulder asserts, this must certainly apply to the collective memory of larger systems as well. The following are a few examples.

On the eve of the new millennium, considerable unrest resulted from the supposed threat posed by the "millennium bug." Deeply embedded in their control systems, many software programs included a date logic that implicitly assumed that the date would begin with the number 19. The transition to the twenty-first century thus implied the risk that all kinds of systems (including those in airplanes and nuclear power stations) would grind to a halt. This situation highlighted the fact that many of these systems were not the product of one programmer, but of whole series of programmers. Many systems proved to be inextricably complex tangles of logical rules produced in time by different programmers, creators, and problem-solvers, which made it very difficult to form a coherent view of them. A similar pattern applies to organizations, which are not constructed, steered, or influenced single-handedly. Organizations display a pattern of learned experiences and the associated sustainable and unchangeable structures.

Memory is also intertwined with the political aspect of sensemaking. In *Guns, Germs, and Steel*, Jared Diamond (2006) shows how the QWERTY keyboard has survived despite the fact that it is an ergonomic nightmare.

> *This book, like probably every other typed document you have ever read, was typed with a QWERTY keyboard, named for the left-most six letters in its upper row. Unbelievable as it may now sound, that keyboard layout was designed in 1873 as a feat of anti-engineering. It employs a whole series of perverse tricks designed to force typists to type as slowly as possible, such as scattering the commonest letters over all keyboard rows and concentrating them on the left side (where right-handed people have to use their weaker hand). The reason behind all of those seemingly counterproductive features is that the typewriters of 1873 jammed if adjacent keys were struck in quick succession, so that manufacturers had to slow down typists. When improvements in typewriters eliminated the problem of jamming, trials in 1932 with an efficiently laid-out keyboard showed that it would let us double our typing speed and reduce our typing effort by 95 percent. But QWERTY keyboards were solidly entrenched by then. The vested interests of hundreds of millions of QWERTY typists, typing teachers, typewriter and computer salespeople, and manufacturers have crushed all moves toward keyboard efficiency for over 60-years.* (p. 248)

3.3.4 Decentralized Memory
In his introduction, Theo Mulder describes how the human motor system consists of a complex whole of relatively autonomous regulatory centers that organize movements in permanent interaction with perceptual, cognitive, and motor processes. It is an example of self-organization within a complex system. As Mulder describes, autonomy and self-regulation are balanced with regulation at the level of the greater whole. This balance can also be observed in the memory of organizations.

Memory is not produced by one central, Big Brother-like consciousness that decides what the world will look like. Memory is a characteristic that belongs to the organization as a whole and which evolves through interaction between its members. It is solidified into a structure, and it is established in procedures, archives, regulations, and agreements concerning values, mores, and habits that support learning experiences. As in the system described by Mulder, however, there is no single, central control system that steers people's actions within an organization. Such a system would be impossible; an organization as a whole is far too complex. An organization controlled in such a centralized way would not be able to retain the necessary flexibility.

> A police-science researcher once told us that the actions of police officers on the street are not determined by the governing authorities, but by citizens. In the Netherlands, the mayor (in the role of police administrator), the chief public prosecutor and the chief commissioner discuss the deployment and priorities of the police force. It is within this triangle that decisions are made

to focus more effort on particular types of problems (e.g., fighting organized crime, enforcing traffic control, or reducing bicycle theft). The chief commissioner translates these decisions to the police organization in the form of instructions, budgets, and similar instruments. Street officers have been thoroughly prepared by an entire catalogue of priorities, tips, and instructions. Nonetheless, should a citizen come running to report that a fight is taking place around the corner, that is where the actual priority will lie.

Locally evolved meaning that is connected to day-to-day actions is more influential in guiding actual behavior than are strategic priorities that are determined at the highest levels. In this example, the professional insights of individual police officers are what ultimately guide their concrete considerations, and not the priorities and instructions that are issued from above.

3.4 The Learned and Learning Organization

This title of this chapter is "The Learned Organization." We obviously selected this term with reference to the concept of the learning organization, which has received considerable attention in recent years. The ability to learn and the ability to develop are important qualities of organizations. It is thus important to understand the relationship between the learned and the learning organization.

The starting point for both the learned and the learning organization is that an organization is a system that has the capacity to learn and adapt; they do not differ on this point. However, there is a difference in emphasis. According to Senge (1992), a learning organization is aimed at organizing metanoia, or continuous renewal. To achieve this, organizations constantly encourage their members to learn and aim for continuous transformation. To achieve a state of continuous learning, a learning organization stimulates the practice of continually questioning assumptions, habits, and unconscious self-evidencies, with the aim of remaining open to new signals. This openness enables an organization to anticipate and adapt to new circumstances time and again, thus remaining successful. The organization is continually "reinventing" itself.

The disadvantage of such a focus is that, in practice, it is likely to result in a lack of appreciation for that which has been learned in the past. Moreover, it involves a risk of underestimating the importance of retaining that which has been learned in the organization's memory. Learning organizations focus on innovations that are required for the future; for them, the adaptive side of knowledge is the more important one. Without intending to detract from the concept of the learning organization, it is also important to realize that knowledge that has been learned is (and will continue to be) present, has proven its usefulness, and is responsible for the organization's success up to this point.

3.5 Organizational Development from Scratch: Knowledge and Knowledge Development

Memory and knowledge played an important role in the Platform against the A6–A9. At the time that the Platform was established, the idea of constructing such a highway in this area already had quite a long history and was connected to a complex whole of facts, opinions, positions, and assumptions that were more or less explicitly formulated. The core members of the group were aided by the fact that relevant knowledge was available within the core group, as well as in the wider network. This included both the explicit knowledge needed to act as an effective opponent to RWS, as well as contextual knowledge concerning the relevant power relations, players, and rules of the game. Available knowledge was stored in the heads of many individuals, and it had been gathered in a variety of contexts. Platform members included urban and regional planners, members of political parties, and active players in the civil society. All of these people were able to make connections between explicit and implicit knowledge, and thus to interpret the memory that had evolved regarding this issue among the parties involved. This helped them to form independent opinions and develop new knowledge with which to arm themselves.

The group gathered new contextual knowledge through interaction with their environment. For example, RWS had an interest in playing the parties involved in the Platform and their supporters off against each other. These attempts provoked the Platform to close their ranks. Once people realized that RWS was using tactics aimed at dividing and conquering, they were able to take action to prevent these tactics from working. An important element of this knowledge concerns power relations, which is the focus of the next chapter.

Contextual and implicit knowledge was developed through the organization's own process of formation. In the process of organizing, the core members became unconsciously competent with regard to internal cooperation. As the evolved way of organizing was repeated more frequently and increasingly successfully, people became more attuned to each other, resulting in knowledge about how to work together effectively and efficiently.

3.6 Summary

The cognitive aspect of robustness entails the capacity to anchor that which has been learned within the memory of an organization. It comprises consolidated knowledge and the capacity to use that knowledge as a point of reference for observations. Instead of a central storage unit that can be accessed and known in a simple way, memory is more like a complex network of history, experiences, and knowledge, and one that involves multiple "administrators."

Knowledge is stored in the habitual actions and routines that evolve over time, in archives, systems, and databases, and in the brains of individual people. These characteristics also enable the cognitive aspect of robustness to contribute to the unchangeability of organizations. Finally, it is harder to unlearn something than it is to learn it in the first place. This is particularly true of unlearning things when we are not sure how they actually work.

4

Power:
The Political Aspect
of Robustness

Another important factor in the unchangeability of organizations is that of stable patterns of power. Power is a sensitive issue for many people. That is why observing the animal world can help us to gain a better understanding of the purpose and function of power and discuss the subject more readily. We therefore turned to ethologist Jan van Hooff to give us an insight into the role of power relations in the animal kingdom.

Power and Stability in the Social Organization of Animals

Prof. Jan van Hooff
Emeritus Professor of Ethology, Utrecht University

In 1922, the Norwegian researcher Schjelderup-Ebbe conducted an experiment with chickens. He brought together a number of hens that were not familiar with each other in a single space. Pecking fights immediately broke out among the chickens. After a short while, however, Schjelderup-Ebbe observed a change. At the beginning of the experiment, any chicken would attack any other chicken. Before long, however, the attacks became one-sided. α attacked β, γ and δ; β pecked γ and δ but did not even dare to think about pecking α. At the end of the series, the proverbial omega (ω) was pecked by all the others, but did not dare to peck back at anyone. A hierarchy of dominance or "pecking order" emerged. This term has become standard in everyday speech, although speakers may not be aware of its origins.

Another remarkable observation was that the chicken fights became less frequent and less intense over time. This was advantageous for both the subordinate and the dominant chickens. By yielding to the dominant chickens, subordinates avoided disastrous confrontations. As this gave the dominant chickens control over the access to resources, it was no longer necessary for them to reconfirm their positions of power continuously.

The struggle for power positions is a universal phenomenon in many different animal societies. It is well known that physical and mental traits that provide an advantage in the struggle for power (e.g., assertiveness, lust for power) are subject to positive evolutionary selection pressure. Individuals for whom these traits have provided access to better territories, better food, better nesting sites, better (i.e., preferred) sexual partners, and other superior resources are thus better situated to produce offspring. This transmits the natural propensity for the traits that have made them successful to subsequent generations.

Power refers to the ability to predict and control situations, whether social or non-social. It implies a welcome reduction of uncertainty. An experiment with rats provides an illustration. Three rats were placed in separate cages for a certain period of time each day. Two of the animals received mild electric shocks at unpredictable times, but simultaneously. The third was the control rat, whose treatment consisted only of being placed in another cage. For one of the rats that received the shocks, a light flashed five seconds before the shock was administered. It was thus a signal for the shock. After some time, the three rats were examined for signs of stress (e.g., gastric or intestinal ulcers). The control rat showed no evidence of ill effects from the treatment. In contrast, the two rats that had been shocked displayed clear symptoms of stress. Remarkably, the rat that had not received a signal for the shocks displayed the strongest symptoms. When this rat was placed in the shock cage, it was in a constant state of anxious expectation, while the other rat became tense only when the light flashed. The same unpleasant situation was therefore more tolerable for the animal that was able to prepare for it.

In another experiment, three animals in the shock cage were given wheels that they could turn. Once again, two of the rats were shocked at unpredictable times. This time, a light flashed for both of them. For both rats, the light was followed by a shock five seconds later. In this case, however, one of the animals was able to prevent the shock by turning the wheel as soon as the light flashed. In time, the rat forgot about it, and shocks were administered to both animals again. Both rats therefore received the same number of shocks. The rat that was able to control the administration of the shock by turning the wheel displayed hardly any symptoms of stress. In contrast to its counterpart, this rat had been in control of the situation.

The predictability of situations and the associated ability to control them are matters that animals strive to achieve. Moreover, dominant animals occupy a central position within the "structure of attention." Other animals notice them and direct their behavior toward them. The powerful invest in "putting on a show"; they engage in impressive behavior that continually draws attention toward themselves. In her population of chimpanzees, Jane Goodall observed how a male discovered that he could increase in status and power through the commotion he created by drumming loudly on an empty gasoline can.

Individuals with power exercise decisive influence, and the others have only to position themselves in relation to the ones in power. As long as those in power follow a consistent line, however, subordinates also have a high degree of predictability, which they can exploit in ways appropriate to their position. Subordinates that have no prospect of achieving a higher position of power have an interest in reinforcing a stable power structure within which they have found their place.

The ability to be in charge instead of playing follow-the-leader neverthe-less remains a tempting prospect for subordinates. Attempts to seize power are therefore regularly observed among social groups of animals. The ways in which this takes place differs across species and according to circumstances. In communities with complex social structures, however necessary they may be, physical force and mental persistence and assertiveness, however neces-sary they may be, are not always sufficient to gain and hold on to power. A world-famous study of the chimpanzee colony at the Burgers' Zoo in Arnhem provided an illustrative example. Although the males in a chimpanzee group form a cohesive block against the males from other groups, they compete for positions of power within their own group. An individual male, however, can never assume a position of power solely through his own strength. This is pos-sible only with the support of other group members. It is therefore necessary to form coalitions, not only with other males, but also with influential females. For those other males, however, the question is, "Why should I support ano-ther group member who is actually my rival?" or "Which of my counterparts do I want to help put in power?" The animals apparently consider two issues (according to intuitive estimation rather than rational deliberation). The first issue involves the question, "Would I have any chance myself?" If the answer to this question is negative, the following question is, "Who would do the most to gain my support and who is most likely to gain power because of my support?" The study showed that chimpanzees aspiring to power tend to "buy" the support of others with indulgence, assistance, and favors. For example, the leader may allow his coalition partner to mate with a potentially fertile female. A supporter who receives too little in return for his support will sometimes change sides.

The role of adult females is interesting. In the wild, females usually refrain from getting involved in the power games of the males. In comparison, the females in the captive population in Arnhem, where the animals spend much more time in each other's presence, are more involved. The females support the leading male (unless he exhibits excessively despotic behavior). In several power shifts, they immediately took the side of the winner. Their interests and the interests of their offspring are best served by peaceful conditions in their habitat. This is achieved by reinforcing the position of the one in power. In short, the quest for power and control is universal.

Violence can be used as a means of seizing privileges. Nonetheless, power is obtained not only through conquest; it can also be granted as an outcome of social transactions between parties within a system of reciprocity. Power can also be based on authority. As observed in a number of species, decisions regarding group behavior (e.g., the daily marching route: where the group will seek food, drink, and rest) are made by certain members of the group. One well-researched example involves Hamadryas baboons. These animals live in large troops on the parched plateaus of Ethiopia and Somalia, where water and food are scarce. Such troops consist of a number of small harem groups, each of which comprises one adult male and one or more females, along with their offspring. Every morning before the harems leave the common sleeping sites to spread out in their own directions on their long daily food hunts, a type of voting process takes place regarding where they will meet each other around noon: near a watering hole. They prefer to drink together, as predators sometimes lie in wait around watering holes. Early in the morning, several males gather in a small assembly. One of them walks in a particular direction and then looks back at the others. As the rest remain seated, a second male stands up and walks a short distance in another direction. The process can be repeated several times. When one male is followed by the others, the group disbands, and the various harem groups go their own ways. The Swiss researcher Stolba, who observed this, decided to walk in the last-indicated direction. Upon arriving at a watering hole along this route, he waited. Around noon, he was surprised to see the same harem groups assembling. He concluded that the animals had met that morning to call for suggestions for places to drink, and that one of the proposed locations had "won." He also found that the older, experienced males often determined the direction, while the others tended to "go along." These older males, however, often were no longer the most dominant. Age and experience thus generate authority.

A similar phenomenon has been observed among elephants. These animals live in matriarchal communities, and it is the older females who make the decisions. These individuals thus have a form of power that does not emerge from the desire of individual animals to seize coveted resources for themselves.

Since certain social structures provide certainty, it is in the interest of those in power and those who provide direction, as well as subordinates to reinforce these structures, although, of course, everything is relative. Those who do not have power are continually exposed to the temptation to shake up the existing order. Throw the bums out! Those who have power, however, are extraordinarily attached to their privileges, and they will relinquish their dominant positions only under strong and irresistible force. The intensity of occasional rank conflicts among animals demonstrates this. No higher-ranking individual will easily relinquish a preferred position in the access to resources; they do not part willingly with the ability to arrange things according to their own preferences, thus securing a position in the center of the attention structure, that is, prestige.

4 Power: The Political Aspect of Robustness

In our work and in "the outside world," the struggle for power can be observed everywhere. In organizations, we see many battles waged about who should take particular decisions. Although some of these battles are overt, many more are veiled. Furthermore, what is just as much at stake in many debates as the particular topic under discussion is the positioning of those participating in the debate. Frequently winning substantive arguments increases the status and strengthens the position of a participant. The patterns that Jan van Hooff describes with regard to monkey games are all too familiar among humans as well. The book titled *Chimpanzee Politics* by Frans de Waal (1991), a student of Jan van Hooff, provides many an example. You cannot read the book without drawing parallels between what the author describes with regard to monkey behavior and the human behavior that one encounters on a daily basis.

The political aspect of robustness involves power, positions, and relationships with regard to each other. As in the case of the aforementioned aspects, this aspect also has both a stable and a dynamic side. We refer to the stable aspect as the pattern of power, and we refer to the dynamic aspect as the power game.

Table 4.1   Aspects of Robustness: The Political Aspect

Facet \ Aspect	Social	Cognitive	Political
Dynamic	Interaction	Learning	Power game
Stable	Routines	Memory	Power pattern

As described by Jan van Hooff, the power game is played out on a field according to its own rules. Although the playing field exists in the present, it has been formed over time. We refer to the playing field that has developed as the pattern of power. It is because the pecking order constitutes a pattern that it is persistent; the omega chicken will always assert its status whenever a weaker chicken comes along.

4.1   The Power Game

The power game is the dynamic side of the political aspect of robustness. In the preceding chapters, the primary emphasis was on how the convergence of images of reality and patterns of behavior generate cognitive and social communality. In contrast, the political aspect organizes difference.

Sensemaking processes do not produce a single, uniform meaning. The formation of multiple social-cognitive configurations (Bolk in Van Dongen, 1996) includes the organization of differences in meaning. These differences form an important engine for the dynamic side of robustness.

Creating realities with the help of sensemaking processes involves more than simply creating new meanings; it also involves the calibration and recalibration of different meanings in relation to each other. When people have differing views of reality (and of the associated differences in behavioral patterns), such differences converge through interaction.

The social-cognitive configurations and islands of meaning described in the previous chapters exist because people associate themselves (as a group) with views of reality. In addition to involving the question of how to make different meanings converge, interaction thus involves the determination of how meanings relate to each other. A game involving the positioning of meanings emerges. The game of positioning meanings is accompanied by a game of positioning sense-makers, who are the actors in the political game. The function of this game is to use differences to clarify relationships to and in comparison with each other.

According to Watzlawick (1990), we assign meaning to objects as well as relationships. Objects are reasonably constant, and their characteristics can be investigated. Differences in the meanings that are assigned to objects can be readily resolved through objective analysis. Ambiguity is easily reduced. The same is not true for relationships and definitions of reality, which are neither objective nor unambiguous.

> Imagine that you and your manager have a difference of opinion regarding your salary. It is relatively easy to reach consensus about the relationship between the job structure and the salary scale. If you are of the opinion that your salary is not proportional to your efforts, however, it will be more difficult to create convergence in the views that you and your manager have of reality. The manager's decision does not feel like an objective decision; it feels more like a "proposed relationship" with regard to you, your efforts, and your person. The difference between your perspective and that of your manager involves more than just the object of your salary; it also touches directly on your relationship. The meaning that is assigned to a relationship is always intersubjective, and it can therefore never be established with complete certainty. In interaction, it is always under negotiation.

4.1.1 Interests

People do not enter into ties with everyone. The interests of people and the ways in which they are connected to configurations play a role in the game of positioning as well. They weigh whether doing so would yield advantages in resources or position, or whether it would yield support for their convictions and views of reality. If such advantages are no longer there, people leave the configuration or the island of meaning and join forces with others. Termeer (2001) describes how configurations disintegrate and change when new people enter the scene with different definitions of reality.

This is possible because people never belong entirely or exclusively to one configuration. People are partially included (Weick, 1979); they belong to multiple configurations or clusters of islands of meaning. Belonging to more than one makes it possible to move across configurations. Van Dongen et al. (1996) argue that people are connected to various configurations through multiple inclusions. They also point out that this multiple inclusion contributes to the differences that exist and to negotiations about these differences.

The processes described previously give rise to an ongoing political game that involves configurations taking up positions in relation to each other and people taking up positions within and between configurations. This is a game that never ends, and it is kept alive through dynamic sensemaking processes. Van Dongen et al. consider difference as a vital factor in sensemaking and for an organization's dynamics. They see the power game and the conflicts that are associated with it as unavoidable within a multifaceted reality. The power game, tensions arising from difference, and conflict are all part of it.

The power game is not without boundaries. The pattern of power that was once built through sensemaking marks out the playing field upon which the power game is played.

4.2 Patterns of Power

It is fascinating to consider how a mere fifteen years after the Bastille was stormed and the French Revolution sought to bring an end to feudal relationships, Napoleon had himself crowned emperor. You could say that, in outline, contemporary French presidents still behave like the French kings of old, with their sovereign power, allure, and grandeur, and a legacy of monuments and behavior that is on the fringes of law and morality, with mistresses and affairs. It is as if the drastic reorganization that was the French Revolution failed to affect some elements in the pattern of power. Elements of existing power relations appear unaffected by other major revolutions as well. For example, despite the Russian Revolution, *glasnost*, and *perestroika*, Putin has a tsarist air about him. Chinese leaders continue to display traits of the inviolable Mandarins, even despite a *cultural* revolution. In *Wild Swans*, Jung Chang (1994) describes the fate of three generations of ordinary Chinese women. Between the lines, readers can see a clear image of a world that is changing drastically on the outside, while the power structure and feudal relationships remain largely stable on the inside.

Closer to home, Dutch leaders are expected to deny their power and behave "normally." In a book entitled *Amsterdam: The Brief Life of a City*, Mak (1999) observes that, for centuries, showing off has been considered inappropriate within Dutch society, which is dominated by the figures of the pastor and the merchant. In the Netherlands, therefore, people with power behave in ways that are egalitarian—working together with others and acting like ordinary people. This makes them no less powerful.

We can also observe similar stability in patterns of power within organizations. Regardless of how much may change, the pattern of power appears to remain the same, and it therefore calls for an appropriate dynamic. Even in the face of major changes, in which the rules and the mutual relationships are considerably shaken up, the underlying pattern appears so stable and strong that it invokes the "old" behavior, and people fall back into old "roles." Historical patterns remain visible, even when organizations have consciously devoted considerable attention to adopting a new style of management.

In an organization with an extremely complex and largely ineffective pattern of governance, it was decided that the board of directors should focus on governing instead of managing processes. In order to realize this change, an extra layer of management was added to the governance structure. This layer was intended to direct the operational processes, thereby allowing the board the space they needed to govern the organization. For a time, it appeared that the desired effect had been achieved. When the board changed, however, the new board rediscovered the old interest in operational, concrete matters. One of the reasons was that this arrangement made it easier to exercise influence. It made the board more visible in the power game than did the practice of governing.

In some cases, patterns of power are anchored in the nature of the organization. The power game within the administrative organization sometimes resembles the game of their political bosses: behavior directed toward spectacle, individual achievement, negotiating behavior, and opportunism.

In hospitals, law firms, research institutes, and other professional organizations, managers are taken seriously if they are vocational colleagues. In a dissertation entitled *De medicus maatgevend, over leiderschap en habitus* (The Physician as Leader: On Leadership and Habitus), Yolande Witman (2008) argues that physicians can have authority within their groups precisely because of the fact that they are colleagues in the same profession. In their capacity as supervisors, they must always remain physicians and colleagues. This allows them to influence their own colleagues. From within this role, they can build a bridge toward the world of management. Witman warns doctors not to start acting like managers through such actions as constantly emphasizing production figures, as such behavior undermines their authority among their colleagues.

In a family business, traces of the fundamental pattern that was started by the founder can remain visible in the decision-making and power structures for years after the company is no longer owned by the family. In such cases, it is said that the culture of the founder can still be observed.

A crisis manager was once a supervisor in a candy factory that was on the brink of bankruptcy. For as long as anyone could remember, there had always been a selection of the company's products on the table in the boardroom. While working long hours with a number of colleagues in order to steer the company away from the precipice, the crisis manager nonchalantly popped a chocolate drop into his mouth. "Too sweet," he said, grimacing, as he took it out of his mouth and threw it into the trash without a thought. Two days later, he was walking through the bonbon factory. Production was at a standstill, and the machines were being converted. When asked, the floor supervisor explained that he had ordered the change. "The candies were too sweet, weren't they? We changed the recipe."

4.3   Patterns of Power: Layered Complexity

Like routines and memory, patterns of power are characterized by layered complexity. They contain meaning and images that provide direction for behavior and that involve the positions of parties with regard to each other, as well as the extent to which some meanings "are truer than others." This pattern of power is a component of the robust fabric, and it is layered.

Clegg (1989) distinguishes a number of levels that can help explain how the pattern of power is built and how its various parts fit together:

1. The first level is the level of *direct* interaction. On this level, sensemaking takes place between people, who jointly determine through interaction how they will relate to each other, who will have more power, and in what way. On this level, the means of power are applied to determine the pecking order. This is the level of concrete behavior.
2. The second level is the level of *rules and procedures*. On the first level, people determined their relationship to each other. On this level, these relationships are established in routines, patterns of behavior, and, more concretely, in rules and procedures. This level describes how people will (or should) behave toward each other in terms of power. This level is comparable to that of regulation, in which formal rules limit behavioral possibilities.
3. The third level is the level of the *power structure*. At this level, power is translated to the way in which the organization is arranged, and it acquires reinforcement within this arrangement.

The third level is the level of the pattern of power; it is the change-resistant "playing board" upon which the power game is played. Although people no longer have influence over it, it does provide direction for how the game is to be played. As a metaphor, it is comparable to how we deal with the constitution. These are meta-rules that are beyond discussion and to which the regulators must adhere. In this way, the constitution has a certain element of inviolability. Proposals that are contrary to the constitution are nearly equivalent to sacrilege. Whereas ordinary laws can be changed with some effort, the constitution can be changed only with large majorities before and after elections.

4.4 Limited Force of Governance Through Patterns of Power

The most complicated feature of the third level is that it no longer involves any administrator. The first two levels can still be "managed" by administrators and managers; the struggle for power can be waged, power shifts can take place, and correcting forces can generate a balance of power on these levels as well. In contrast, the third level is such a complex entirety of factors that there is no single figure at the helm. It is an ungovernable bulwark that was once designed by individuals, but which now stands alone. It has become so self-evident that it has taken on a life of its own. It continues to exist because individual employees behave according to the prevailing rules, and these rules subsequently restrict the space available to change the situation. Power thus reproduces and maintains itself.

This is contrary to the tendency of people to ascribe ultimate power to the ruler. After all, this leader is involved with everything and should therefore be capable of changing everything. The history and the construction of the pattern of power suggest a different situation.

In Dutch municipalities, the already complicated game between the board, council, and administrative organizations has not become easier since the passage of the Separation of Powers Act. In some cases, major contrasts are experienced, and cooperation often proves difficult. As the chair of both organizations, the mayor has a formal, albeit limited, power position. Nonetheless, the mayor's authoritative role does lead to the situation in which he often has access to power. The message that the mayor receives from both the board and the council is often, "Make sure that our relationship remains favorable and make sure that we are all effective." The message behind this message is often, "Keep that other party off my back and make sure that they become smarter, more efficient, more customer-oriented et cetera." The mayor can fulfill the first role as a networker, but not the second. The mayor is not the figure who has that authority.

For a variety of reasons, formal rulers are not always the ones who are actually in charge. For example, the primate researcher Frans de Waal (2005) once outlined how President Bush may indeed have been the "king" among his people, but that he held that status in part because the alpha male and "king maker" Dick Cheney pushed him forward.

In the Netherlands, it is easy to see how those in power are dependent upon each other. The country has a tradition of forming coalitions. This was already the case during the reclamation of land, which would have been impossible without cooperation among neighbors. This is reflected in the national government, in which no single party ever has a majority. It can also be observed in the way in which the top management of organizations often consists of a collective. Boards of directors, advisory boards, and commissariats consist of people who must work matters out as a group, but who also must guide the interplay with the other boards in the organization.

No monarch is without a court. The British television series *Yes Minister* provides a wonderful illustration of how a supporting official can actually be the ruler. In organizations, there are entire networks of formal and informal power that place and maintain the responsible party in the position to be able to bear such responsibility.

Those with power are often restricted in organizations as well. Staff members, advisors, and secretaries determine which information reaches the boss and which information the boss does not need to see. This places those at the top in a position that is insulated from the rest of the organization. In this way, those with less power contribute to the span of influence held by those with more power.

Finally, those who have power granted by legislation are embedded within an even greater pattern of power. Joint decision-making legislation is an example in which formal power relations between administrators and employees (and thus between those with less power and those with more power) are "arranged" within an organization. The collective labor agreements, the Works Councils Act, or agreements proceeding from the administrative consultation committee constitute the rules of the game. Although the administrator and the works council do spar with each other, no joint decision-making organ would dare to stage a coup outside of these rules. Why should they? Everything is arranged so well.

> During a meeting of the directorate, a proposal from the personnel department to adopt a regulation to provide ergonomic equipment was under discussion. As the memorandum was being read, it became clear that the proposal involved the provision of VDU (Visual Display Unit) spectacles. Although the interim director of this organization was opposed to the proposal, it proved impossible to do anything but agree. The occupational safety regulations and the role interpretations of the personnel officer involved were more compelling than the ultimate responsibility of the interim director. Not agreeing would cause considerable upheaval, and the proposal was not important enough to warrant this.

4.5   The Importance of a Definite Reality

The reality that people experience is always realized through a process of sense-making. In addition, there are always people who leave a stronger or weaker impression on the dominant meaning. Those whose interpretations leave a strong mark on the definition of reality have an interest in keeping this reality as it is. Homan (2006) describes such owners of interpretations as "regime guards." They protect the interpretation and observations against anything that would undermine their view of reality. One of us remembers once having conceived a brilliant idea for a course. He was on tenterhooks to protect his own brilliant idea, which made him listen intently to the arguments made by anyone who wondered if it could be done another way.

Those who have not managed to make a strong impression also assume a place in the pattern of power. Groups who wish to exercise their influence despite their relative powerlessness move toward a local island of meaning in the organization from which they let their voice be heard. For example, groups that were opposed to a merger that took place years ago may continue to air their discontent from within their local island of meaning. This voice no longer has any influence on the most dominant meaning. The members of this group find their confirmation, as well as their identity, largely in each other.

4.6 Roles: The Personal "Template" in the Collective Structure

The seventeenth-century Dutch playwright Joost van den Vondel wrote, *"The world's a stage: Each plays his role and gets his just reward"* (in Skrine, p.4). Roles are an important mechanism for preserving the power structure and patterns of power. The roles that individuals receive define the manner in which they are expected to interpret the social order. Roles are not linked to people; they are part of the social order, which requires interpretation—not only in terms of tasks and substance, but also in relation to others. In this respect, roles are not voluntary. They bring order within a group; they regulate expectations concerning individuals' own behavior as well as the behavior of others. They also make the social reality and the behavior of people within it predictable. Roles also contribute to the identity of those who hold them. Roles thus create clarity for the actors who play them, as well as for the people in their surroundings.

Goffman (1956) describes how roles become institutionalized and how they evoke expectations through social interaction. The role becomes a collective representation of which both the actor fulfilling the role and the people surrounding the actor have expectations regarding the nature of the behavior. Roles are conventions. Goffman describes how both actors and observers enact the play of this collective representation, meet each others' expectations and even "look away" when behavior is displayed that is not completely consistent with the role. The social order uses roles to arrange which behavior is to be displayed and seen. In this way, roles are the templates in which individual players carry out the expectations of the power system—in part, because this provides them with their own social identities and in part, because it is expected by the group.

This fact provides deeper insight into why even the most powerful actors are relatively powerless with regard to changing the pattern of power. Often they are not the ones who create the power game, and the game does not exist for their sake. Regardless of their motives, the most powerful actors are also simply players in the game, who are restricted in their direct interaction by the rules contained within the power structure. This structure is a construct that is created; all members of the organization move within this construction of reality, which is accepted as true. Particular behaviors are expected of those with power. They are also not free to step outside of their roles, to behave as if they are powerless or to assign power elsewhere. Such behavior is not appropriate to the reality that has been created.

4.7   Individual Behavior: Freedom in Restraint

If roles and role expectations do not make people remain within the pattern of power, group norms and group pressure exist to ensure that anyone who seeks to change the rules of the game is "reined in." The social influence of the group plays a major role in this regard. The pressure of the group on individual members to adhere to the group's rules is great, and the penalties are harsh. Exclusion, being assigned to a low position in the pecking order, and a lack of appreciation are resources that a group can apply in order to make it clear to individuals that they are not behaving according to expectations.

The more dependent people are on each other, the greater is the pressure to conform (Brown, 2006). Partially for this reason, people have a tendency to go along with the group norm and set aside their own opinions.

> In conflict situations, people express themselves differently in individual interviews than they do in the group. In a study of a conflict between a department head and employees, the employees reported in interviews that they did not really have a problem with their supervisor. Although they may have said he was not the easiest of people, individual employees reported that they did not consider him to be "as bad as the others do," often expressing that they "were quite capable of seeing eye to eye with him." When asked if they would be willing to state this view in the group, however, they responded that they would not. If they were to do so, they would be undermining their colleagues.

Groups apply heavy sanctions for behavior that deviates from the norm, even if such behavior is actually in the interest of the group as a whole. An interesting example in this regard is the classic experiment of Kahn and Boulding (1966). In half of the groups they studied, the researchers placed someone who was trained to act as "the devil's advocate." This person had been assigned to provide critical input in group decision-making process. The groups that had a devil's advocate performed significantly better than did the groups in which no one took this role. Nevertheless, when the groups were assigned to vote one member out of the group, the votes fell without exception on the people who had been decisive in the superior performance.

One reason for adhering to the group norm is that people do not like to fall outside the group. Although the group's experience of reality may be restrictive, it also provides a context within which individuals can understand the world. Moreover, the group and the social order provide individuals with a social identity, and they offer safety and certainty regarding who an individual is. They provide certainty regarding where an individual belongs; it is therefore a source of support and warmth as well. The need that people have to belong thus contributes to the unchangeable character of the pattern of power.

4.8 Organizational Power from Scratch: Playing Along with the Power Game

The subject of the A6–A9 junction takes place in a context with an old and stable pattern of power. Since the Second World War, the Netherlands Directorate-General for Public Works and Water Management (RWS) has been a powerful party in the construction (and reconstruction) of the Netherlands. It has been described as a "state within a state" (Metze, 2009–2010). The pattern of power is determined largely by RWS and its close relationship with the minister and the political field. Other players can exercise far less influence. Furthermore, RWS is by far the most powerful player in terms of money and capacity, and it is in a position to perpetuate its power by manipulating other parties. For example, RWS contracted nearly all of the major engineering and law firms in the course of the process so that none of them could support opposing parties, as they were already involved on behalf of the other party.

Technical and financial possibilities (and impossibilities) play a major role in the power game concerning the construction of a new motorway, as do political relationships and media attention. Although the Platform had originally taken the role of the underdog, it developed into a strong player. To accomplish this, it was necessary for the Platform to gain insight into the pattern of power that prevailed around the subject. They therefore needed to take up a position within this pattern and enter the game in such a way that it could turn the pattern of power in favor of the Platform's own goals.

The media were an important party for the Platform, as they devoted considerable attention to the potential underdog. They were also relatively accessible to the members of the Platform. Local governments (e.g., municipalities and provinces) were also important for the Platform. As these players stood for approximately the same interest, it was largely a matter of building trust and dividing roles, as was also the case with the associated nature and environmental organizations. The possibilities and limitations that these parties had in the game were different from those of the Platform. The political parties were obviously important as well, given that the ultimate decision-making process would take place in the Dutch parliament. The connection naturally worked better for some parties than it did for others. As the area concerned is in the general vicinity of Hilversum, the center of the television industry in the Netherlands, the many celebrities who lived in the area were important potential new players as they were capable of generating considerable attention.

These celebrities ultimately ensured that the Platform developed an advantage in the power game. The phenomenon was completely unfamiliar to RWS, which did not know how to react to this strategy. For example, an official from RWS asked the Platform (in vain) to provide addresses of celebrities, as the minister wished to invite them to a conference.

The power game had to be played within this pattern of power and with the associated players. In this game, the knowledge generated by the Platform proved useful in putting up strong resistance against RWS. The Platform's standpoint was important as a carrier of the public debate and in terms of administrative influence. The public debate concerned not only the A6–A9, but also alternatives. The Platform succeeded in presenting the notion of a motorway that would run through their own vulnerable area in such an unattractive or impossible light that the individual political parties began to speak out against it. The minister had no choice but to join with them and once again start searching for alternatives.

What remained unchanged? The fundamental pattern of the power of political/administrative organization relative to social opponents remained the same. The game itself remained fundamentally unchanged. This group did succeed in becoming a strong player with a different tactic, and this success played a strong role in determining the balance of power in the debate. The motorway was not constructed. The Platform played the power game well.

4.9 Summary

Power is important in organizations. It forms a fundamental pattern that provides direction for the behavior of people. The subject of power is often vested with taboos, however, and it is therefore difficult to grasp in many cases. At times, those who have power wish to deny that they have power. Furthermore, absolute and sovereign power hardly ever exists. This power is distributed across multiple players, and this distribution actually increases robustness. The power game that is played in organizations always takes place within the framework of power relations. This stable pattern of power is so self-evident and surrounded by taboos that it evades change. We have yet to encounter any principals that have ordered a change in these relationships in such a way that they were actually willing to accept less for themselves.

In addition to formal or informal individual power factors, aspects of group dynamics often play a compelling role in the prescription of behavior.

5

Pathological Forms
of Robustness

There are also unhealthy forms of robustness. At the University of Wageningen, Piet Boonekamp has been studying the health of potato plants and threats to their health for years. In this introductory essay, he discusses the roles that changeability and unchangeability play in the struggle for health and survival.

A Chance Meeting Between Phytophthora and the Potato?

Dr Piet Boonekamp
Business Unit Manager at Plant Research International in Wageningen

The Accident of the European Potato-Eater

In Peru and Chile, remnants of potatoes have been found in ruins dating back over 2,000 years, attesting to the fact that potatoes were a staple of the Incan diet. Around the middle of the sixteenth century, the Spanish conquistadors brought the potato back with them from the Andes to Europe. Nobody wanted to eat it, as the entire plant is poisonous, except for the edible tuber. The potato thus did not have a very promising future in Europe. It was only in the late eighteenth century that the potato got a second chance. The threat of food shortages loomed, as the Industrial Revolution had led many agricultural workers to migrate to the cities. Grain production, which had always been sufficient up to that point, had become insufficient to feed the population. The botanists of the day recognized the far greater potential of the potato, and they were supported by a clever marketing strategy on the part of the French king. He commanded his gardeners to grow large quantities of potatoes in the royal gardens. The harvested potatoes were repeatedly presented to the king with great ceremony. Meanwhile, the gardens were guarded day and night. On some occasions, however, the guards were purposely withdrawn during the night. Naturally, the farmers took advantage of these unguarded moments to sneak into the garden and steal potatoes. This strategy helped to popularize the potato and make it the most important source of nutrition in Europe, without ever having had to resort to coercion. Within a short time, a large part of the population had become dependant on the potato. A prime example is that of the impoverished people of Ireland, whose diet came to consist solely of potatoes in the nineteenth century, with each individual consuming around five kilos a day.

An Accidental Meeting Between the Potato and Phytophthora in Europe

The potato plant has its origins in central Mexico. In addition to the cultivated potato species *Solanum tuberosum*, many wild potato varieties can be found in the region. It is also a breeding ground for Phytophthora ("plant-destroying") pathogens, which also show wide diversity. Cultivated potato varieties, which spread from this area across the whole of Central and South America, are especially vulnerable to Phytophthora. In contrast to Mexico, conditions in the

Andes are not always ideally suited to infestation. It is from this region that the Spaniards took potato plants with them, and these plants were not infested with Phytophthora. Was this a coincidence? Probably not: it seems safe to assume that they deliberately selected plants from the healthiest region to take with them. As Phytophthora was not native to Europe, things went quite well for a while. The meeting between the European potato and Phytophthora was not inevitable. In the nineteenth century, however, potatoes began to be imported from Mexico, the primary breeding ground for Phytophthora. In 1845, the disease led to famines across Europe. In Ireland, more than a million people died, and another million emigrated to the United States. Was the disease imported by accident? Probably not: evidence suggests that people knew of this terrible blight and wanted to study it, thereby letting the genie out of the bottle. Was this a pure case of bad luck? No: there was considerable skepticism among the scientists of the day, and a fierce debate was being waged on whether fungi were the result of plant diseases (the commonly held view at that time) or rather the cause (the new line of thought). Phytophthora provided evidence to support the latter theory, which constituted a major contribution to the development of scientific thought. It also meant that Phytophthora had come to stay and that it would remain the foremost potato disease.

The Near-Destruction of the Potato by Phytophthora in 1845

The meeting between Phytophthora and the potato resulted in a battle, in which the main weapons were both species' capacities for change. Phytophthora can survive only by infesting potato plants. To be able to infest a potato plant, the organism must possess a virulence gene. The potato plant can arm itself by possessing a resistance gene. One particular resistance gene can serve to counteract the effects of one particular virulence gene, thereby stopping the Phytophthora infestation in its tracks. In response, Phytophthora must develop a new virulence gene, which can circumvent the potato plant's resistance gene. This is apparently what happened in the case of the Phytophthora variety that was introduced to Europe in 1845. In the space of only a few years, almost all potato crops had been infested. Nonetheless, a few potato varieties survived. These varieties apparently possessed an appropriate resistance gene, and they were used to cultivate a new population of plants that were resistant to Phytophthora.

Why did Phytophthora fail to develop a new virulence gene quickly in order to infect these resistant potato varieties as well? The answer was a lack of sex. Fortunately, only one Phytophthora clone had been transported from Mexico. This clone could reproduce only through spores and not through cross-fertilization, which is an organism's best way of combining its own genes with those of its partner, thus passing on new combinations to its offspring. This would have allowed Phytophthora to acquire a host of new virulence genes from a range of partners in a very short time. The remaining potato plants would not have been able to develop new resistance genes quickly enough and would therefore have had no chance to survive. The lack of sexual partners meant that Phytophthora was forced to revert to the strategy of trying to adapt its virulence gene in such a way that they would render the resistance genes of the new potato varieties useless. This is a slow process of trial and error. Understanding this process requires insight into changes at the level of DNA. A gene consists of a DNA helix made up of 10,000 letters, so to speak, which combine to form one very long code. To retain the meaning of the code (i.e., the gene's function) in an organism's offspring, the string of letters in their genes must remain the same. It is a natural though infrequent process, however, for individual letters to change occasionally. This process is known as "mutation." Genes that have been changed through mutation can be passed on to any offspring an organism might have. In almost all cases, mutation weakens the gene's function (due to the code having become illegible), thus producing a disadvantage for the offspring. In some cases, it can result in a new code with a different meaning. In this way, after all manner of "lesser" genes, the arbitrary process of mutation could eventually produce a new virulence gene. The odds of this happening are low, although it is statistically likely to occur in at least one Phytophthora specimen within a period of 150 years. Such a strain of Phytophthora would consequently have had a highly competitive advantage with regard to the infestation of the remaining potato varieties. It would have completely supplanted the old Phytophthora population in a short space of time, and it would have prevented the potato from having a second chance in Europe after 1845.

This scenario would certainly have happened if the potato growers of the nineteenth and twentieth centuries had not began to grasp the rudiments of how to improve crops. In contrast to Phytophthora, potato plants did have the advantage of cross-fertilization. The scope was rather limited, as specimens could cross themselves only with related cultivated strands, whose gene packages were largely of similar content. Nonetheless, cross-fertilization did produce more opportunities for changes in resistance genes than mutation would have on its own. Because potato plants take years to mature before they produce offspring, however, it is not possible for potato strains to respond to Phytophthora varieties quickly.

The potato plant did have one great advantage on its side, however: potato breeders. They were the agents who selected the best pairs of potato plants for cross-fertilization, instead of leaving it up to chance. They were not able to influence the combinations that occurred in the offspring of various specimens, as these combinations arise arbitrarily. Breeders were nonetheless able to extract those individual specimens from the offspring that were resistant to all common Phytophthora strains through *selection*. Even without knowing anything about genes, breeders managed to breed sufficient new resistance genes into potato plants to keep Phytophthora at bay for almost 150 years.

Why Was There Nonetheless a Near-Disaster in 1980?

The disappointing potato crops in Europe following the hot, dry summer of 1975 led to the import of tons of potatoes from Mexico (instead of from the Andes, as the Spaniards had done four centuries earlier). Along with these imported potatoes came new Phytophthora populations. This had predictable consequences: it was indeed possible for the new population to have sex with the old population. For the first time in 150 years, Phytophthora had the opportunity to acquire new virulence genes through the additional factor of cross-fertilization, instead of merely through mutation. This gave Phytophthora a true advantage in its long-standing stalemate with the potato plant. Within a few years, the old Phytophthora population had indeed been entirely supplanted by a new variety with an entirely different virulence gene. This meant the potatoes cultivated in Europe were suddenly highly vulnerable to Phytophthora. Was this a coincidence? No: people could have known that importing such large quantities of potatoes from the breeding ground of Phytophthora would involve this risk. Could it have been prevented? No: the globalization of trade would have eventually led to the introduction of the new Phytophthora population to Europe at some point. The advantage of this situation was that it served to accelerate efforts to introduce new resistance genes into cultivated potato varieties. The best source of these genes is in the wild potato varieties of central Mexico. Unlike cultivated potatoes, these varieties are never affected by Phytophthora, despite its prevalence in that region. Combined with the experiences of growers, research has generated a variety of techniques for transferring new resistance genes from wild varieties to the cultivated potato. This process is expected to redress the balance. Nonetheless, the battle will continue.

Conclusion

It is often said that organisms such as Phytophthora change in order to increase their chance of surviving amid changing circumstances. This falsely implies some kind of deliberate, systematic approach. All of the changes that occur in Phytophthora are purely accidental, and most prove fatal to the individual plants. Changes are only rarely advantageous, thus enabling the changed individual to take over the population. The emergence of populations that are better adapted to their changing circumstances is thus the result of coincidence, and it happens at the expense of the old population, which becomes extinct.

What is our role in this process? Although we may not be able to steer coincidence, we can steer its results through our interventions. This is what has allowed us to keep the cultivated potato thriving.

5 Pathological Forms of Robustness

The arms race between the potato and Phytophthora, as described by Piet Boonekamp, exists because both are capable of using small adjustments to maintain themselves under changing circumstances. The potato and Phytophthora are both characterized by robustness. They possess stable characteristics, which they protect by adapting their genes for either resistance (in the case of potatoes) or virulence (in the case of Phytophthora). Pathological situations emerge whenever one of the characteristics of robustness becomes dominant. In the case described here, such a situation would mean the end for either the potato or the Phytophthora.

5.1 Overshooting the Mark

Although robustness enables individual entities to retain their identity under changing circumstances, it is not a characteristic that automatically fosters a healthy balance within an organization. If that were the case, organizations would never need to change at all. Robustness can become unbalanced, and it is at these points that we deem change necessary. When the balance shifts too far in one direction, however, robustness can take on pathological forms. When this happens in an organization, it ceases to react to information and change in a normal way.

In pathological situations, organizations are either excessively or insufficiently closed off from their environments. The first situation, which is characterized by a strongly closed character, produces a form of pathology in which an

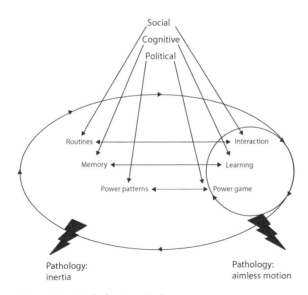

Figure 5.1 Pathologies of robustness.

organization's dynamics (i.e., its capacity for reinforcing robustness through sensemaking) disappear, resulting in inertia. In the second situation, in which the organization is entirely caught up in its environment, there is too little stability, which results in a lack of direction.

Table 5.1   Pathological Forms of Robustness

Facet \ Aspect	Social	Cognitive	Political
Dynamic	Lack of routines and certainty regarding actions	Lack of organizational identity and historical roots	Power taboo
Stable	Routines, rules and procedures come to dominate	Learning becomes one-sided and aimed at problem solving within the existing context Signals are missed	Power obsession

The pathological forms of robustness can also be described in terms of their social, cognitive, and political dimensions. This chapter is structured according to these dimensions.

In the interest of clarity, we describe both types of pathological robustness in their extreme forms. In practice, however, most cases display facets or gradations of pathological robustness.

5.2   Loss of Dynamics

The inert form of robustness is characterized by an excessive emphasis on reinforcing and perpetuating the status quo. In an earlier chapter, we described how sensemaking processes are always aimed at both perpetuating robustness and making adjustments that make it possible to maintain that robustness. In this form of pathology, sensemaking processes are actually used only to perpetuate robustness. If this continues, it leads to inertia and patterns of repetition. Figure 5.2 illustrates how the sensemaking processes strengthen robustness and not change.

The dynamic aspect, which contains the seeds of innovation, increasingly recedes into the background. The organization's primary focus is on existing meaning, routines, habits, and cultures, and it becomes less open to new information and adjustments. The system's capacity for adaptation is lost, and structure becomes dominant. When meanings converge too closely, situations can emerge in which predictability, control, and hierarchy are dominant (De Moor, 2005). The status quo must be maintained, whatever the cost. Organizations

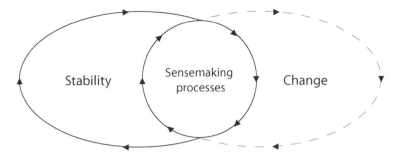

Figure 5.2   Inertia.

fail to react to impulses for change, or they react in ways that are inconsistent with the intentions and expectations of the agents of change.

As long as its environment does not specifically call for a different reaction from an inert organization, inertia will tend to remain invisible. An inert form of pathological robustness will become apparent only when the environment starts to make demands of an organization. If an imbalance develops between stability and dynamism, the organization will no longer be capable of reacting appropriately to the demands of its environment.

5.2.1   *Compulsive Routines*

When inertia develops, routines, rules, and procedures become coercive and almost impossible to change or adjust. In an inert organization, the foremost concern is whether people are adhering to routines, and not the way in which the routines are helping the organization to achieve the goals it has set for itself. Failure to follow the rules is penalized.

Compulsive routines can be an expression of fixations existing in the interactions between individuals. Termeer (2001) distinguishes between social and cognitive fixations. In social fixations, people are no longer prepared to reflect on their mutual relationships. This can lead to ritual behavior, fixed role patterns, and the exclusion of third parties. In these types of situations, behavior is guided by stereotypical views of others. Interaction is fixed and no longer changes.

The extent to which adherence to routines is unhealthy obviously depends on the type of organization. While we may consider it perfectly sensible for procedures to play an important role in such organizations as courts or pharmaceutical companies, we are likely to expect a theater company to display a more ad hoc style of organizing. This example illustrates the context-specific relation between stability and dynamics.

> The phenomenon of compulsive routine can be illustrated by an advertisement in which a mother and her daughter are standing in front of a counter. They have come to claim an inflatable crocodile that the little girl had lost. The crocodile is waiting for them in a corner, in plain view, but they are not

allowed to take the toy with them, as the mother is first required to complete an endless series of forms.

This image is obviously exaggerated. A more realistic example of a situation in which compulsive routine can play a role could be that of a government bureaucracy whose practices are so dominated by procedural conditions (e.g., lawfulness, legal security, and transparency) that initiatives for change intended to increase efficiency or output simply fail to take root.

5.2.2 Closed-Mindedness

Closed-mindedness and inertia lead to one-sidedness in learning. Termeer (2001) uses the term "cognitive fixation" to refer to this situation. Cognitive fixation occurs when people take their own definition of reality as the rule and are no longer able to see other definitions of reality. Van Dijk (1989) distinguishes three types of fixation:

• Empirical: the definition of the current situation is fixed.
• Utopian: there is a fixed, idealized vision of the future.
• Instrumental: there is a fixed causal relation between the past, the instruments to be applied, and the future.

These cognitive fixations prevent people from looking at the world around them in a different way. Moreover, as they are caught up in their own fixations, they are no longer able to look at things from another point of view. The result is an increase in single-loop learning (Argyris, 1991), which involves applying familiar problem-solving routines. Double-loop learning, which involves reflecting on one's own behavior, on the chosen routines, and on underlying assumptions, occurs rarely, if at all. This one-sided orientation and closed-mindedness also causes organizations to miss the signals emanating from their environments.

Consider the example of the Dutch paper manufacturer Van Gelder Papier, a prominent company that went bankrupt in 1981 after being in business for three centuries. The sociologist Cees Lescuere conducted a study of the company's history and the way in which it collapsed. According to this analysis, the power and pride of the company had reached such a level that, in its last decades, they caused the company to miss a number of signs that things were not going well. The level of self-confidence within the company was so high that, although signs that indicated failure, growing competition, and new technologies were heard and were noted down, their significance was underestimated. The organization ultimately collapsed because of its own excellence and arrogance.

5.2.3 Power Obsession

When an imbalance emerges in the political aspects of an organization, it generates an obsession with power. In such a power-obsessed organization, discussions primarily revolve around who takes what decisions. In its healthier form, power helps an organization to organize itself. In a power-obsessed organization, the focus has shifted away from power as a means through which to achieve organizational goals toward power as an end in itself.

In this kind of pathological situation, sensemaking revolves solely around the confirmation of the existing pattern of power. All activities are oriented toward confirming or negating that pattern.

> Consider the example of one student committee at the Academy for Social Work in the Netherlands in the 1970s. This committee consisted of two study groups, one focusing on academic content and the other on democratization. Operating largely independently of each other, the first study group worked hard at politicizing the contents of the academic program, while the other made fervent (and sometimes successful) attempts to reorder the relationships between students and lecturers. The spirit of the times dictated that educational program should revolve around the student. Democratization became an end in itself. Power was no longer a means of achieving tangible goals; it had also become an end in itself. Although the committee's activities were quite instructive for politically active students, they left little room for normal lectures or study groups. The primary objective of education no longer played a central role.

The strongest complicating factor of this form of pathological robustness is the fact that every attempt to change is seen as a political act, and it thus becomes part of the game. There is no longer a party that can regulate this process. In the preceding example, any questions that the academy's directors may have raised concerning the usefulness of the student committee's activities would have immediately been interpreted as an attempt to isolate the committee. Those who raised such questions would have called suspicion upon themselves, thereby providing the student organization with even more reason to increase its efforts against the authoritarian administration. The alternative would have been to take a more understanding line. This tactic was accompanied by its own risks, however: those choosing this path were likely to be accused of "repressive tolerance."

5.3   Loss of Stability

Another form of pathology arises when sensemaking processes fail to generate sufficient guiding constructs and behavioral patterns. In the process of organizing, organizations are unable to establish meaning sufficiently and render it permanent. This results in a divergence of meaning (De Moor, 2005). Organizations lack (or lose) the ability to anchor and retain the structure effectively enough to ensure continuity. Everything seems to be open to discussion, and nothing can be taken at face value. The sensemaking process continuously produces new meanings that fail to become anchored in robustness.

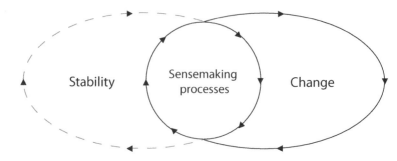

Figure 5.3   Aimless motion.

In Figure 5.3, sensemaking processes are aimed solely at change. Insufficient attention is paid to robustness, and this produces aimless motion.

This pathology is not an expression of unchangeability. On the contrary, the organization's reaction to its environment is one of unguided, unpredictable change. We have nonetheless decided to include this type of pathology here, as it is characterized by an extreme lack of robustness.

5.3.1   When Routines Are Taboo

When a movement toward dynamism gets out of control, it leads to a lack of stability and certainty. This pathology produces an organization with an almost complete lack of routines. The only routine to which people adhere is that of avoiding routines. People in such organizations cannot be sure of the situations in which they find themselves. They can never be sure whether they will find the same situation when they come in for work the next day, as someone may have decided that things should be changed around in the mean time. This lack of certainty causes people to retreat to the level and sphere of influence over which they do have control. In many cases, these areas are limited to their formal positions or duties. Connectivity within the organization is lost, and people make decisions based on their own individual sensemaking processes. This results in a lack of collective memory and repertoire of action.

5.3.2   When Memory Is Taboo

A lack of stability also leads to a lack of attention to the importance of memory. It becomes taboo for people to retain and cherish what they have learned. In their aim to reflect the state of the art at all times, organizations go too far in adopting the latest systems and management methods. This makes people quick to equate existing knowledge with outdated knowledge. That which was learned in the past is set aside as old-fashioned and conservative. In such an organization, people are afraid to speak up and say that something has been tried before, as this would indicate a lack of the desired drive for renewal on their part. This taboo ensures that lessons learned from earlier experiences are rejected on grounds of general beliefs and preconceptions.

5.3.3 When Power is Taboo

Some organizations dare not contemplate the subject of power. Power is equated with the abuse of power and therefore condemned. Not only is power carefully avoided, its existence is even denied or described in euphemistic terms. An illustration on a calendar by the Dutch artist Peter van Straaten shows a clearly well-to-do gentleman in a flashy car telling a younger woman, "I don't have power, honey, I have influence."

The notion that power might be a healthy and necessary characteristic of organizations is unthinkable in these kinds of organizations. People who have power, even if it is derived from their positions, reject it and say that they prefer to be led by other factors, such as initiative and the needs of staff members. In this kind of situation, it is easy to lose track of who is responsible for what, and nobody is left who can take decisive action. Everyone has a say on everything, and anyone can endlessly restart discussions about things that seemed to have been settled already.

> A large government organization decided to introduce the principle of "collegial" management. The principle was anchored in the organizational structure in the following way: group decision-making was favored over hierarchical command structures. Supervisors were selected according to their acceptance of the taboo regarding power games. This was consistent with the vision that anything starting from the grassroots level was good by definition. As a result, the employees started to manage their own individual patches of work. Any sense of a cohesive whole had disappeared, having been replaced by a cacophony of actions and messages that were no longer consistent with each other. Another consequence was that, whenever decisions actually were made, there were always groups of people who felt that the decisions did not apply to them or that they had the right to question those decisions. This deepened the lack of cohesion and prevented people from making decisions that could be perceived as directive. This left staff members with only one option: to keep doing their own work as they saw fit. As a result, people were doing all sorts of things, with no central focus to the activities and no decisions and no choices being made. In this way, the decision not to organize power gave rise to powerlessness.

5.3.4 Cosmology Episode

A specific form of aimless motion arises when robustness suddenly becomes unreliable. This situation occurs during crises. In these situations, the usual behavioral patterns no longer apply to the situations in which people find themselves. People no longer feel that they are part of a rational, orderly, and intelligible environment. Weick uses the term "cosmology episode" to refer to this type of situation (1993a). In a cosmology episode, people are thrown back on their own devices; they start to act as individuals and no longer as members of a larger, cohesive whole. The structure that holds the group together collapses.

In several articles analyzing disasters, Weick describes the way in which a cosmology episode develops (1988, 1990, 1993, 1993a). As long as a disaster unfolds in a more or less predictable way, teams act according to the standard operating procedures. Feedback from the environment helps you to react and bring the disaster under control. In situations in which the environment provides unexpected feedback on people's actions, a number of things happen. For one, it takes a while before people realize what is happening. At first, people fail to recognize the new information. Once it is noticed, people have only their standard repertoires to which they can revert. Although this repertoire is applied, it is no longer effective. People are unable to generate new meaning with regard to what is happening, and they no longer have recourse to appropriate behavior. At this point, the team falls apart, and its members start to act according to their own understandings and their own meanings. Reality becomes incomprehensible, control is lost, teams fall apart, and people are left to their own devices. As a result, the crisis escalates to become a true disaster.

Fortunately, we hardly ever experience such situations. This is partly because, in uncertain times, people look for certainty wherever it can be found. When a situation is characterized by aimless motion, certainty is lost and people tend to hold on to the things they feel they can trust. As in the preceding description, this often consists of people's own circumscribed duties and their own little teams. When organizations move aimlessly, people become incapable of "heedful interrelating." In his works, Weick discusses this concept. People act "heedfully" if they "act more or less carefully, critically, consistently, purposefully, attentively, studiously, vigilantly, conscientiously, pertinaciously" (Ryle in Weick and Roberts 1993, p. 361). When people act in a heedful way, they base their actions not just on routine, but on awareness and active attention as well. Heedful action combines thought, feelings, and desires. Heedful interrelating safeguards organization and learning within the robustness of which people are aware.

5.4 Neglected Organizations

Recent studies have begun to consider a specific type of unhealthy organization: the neglected organization (Kampen and Schuiling, 2005). The neglected organization shares certain characteristics with both forms of pathology described previously.

Neglected organizations no longer react in ways that theories of change would lead us to expect. Changes repeatedly fail to catch on, and it seems increasingly difficult to implement any kind of change. In this type of organization, repeated changes (e.g., reorganizations, mergers) have given rise to a kind of chronic fatigue in response to change initiatives. The entire system is armed against imposed changes, and it tries to survive within an unpredictable environment.

The internal connections have become so disrupted that the organization falls apart into isolated units that operate autonomously, solely according to their own interests. This is because the employees of such organizations have been swamped by repeated series of changes that have not been anchored properly and because due attention has not been paid to supporting people throughout the processes of change.

People are no longer motivated by a sense of attachment; the organization is no longer dear to them. Neglected organizations are characterized by broken relationships and people acting in isolation. They no longer attune their actions to the system as a whole, acting solely upon their own interests.

In our view, these types of situations involve aimless motion at the top, which has the contrary effect of strengthening the sense of attachment people in the workplace feel to the familiar robustness that they have constructed. The organization's highest layer of management is attuned to signals from the outside, and it chooses (or is forced) to react to these signals with ad hoc changes. Such changes are more focused on placating the outside world than they are anchored in day-to-day working processes. People in the workplace develop survival strategies that are aimed at conservation. These strategies thus share a number of characteristics with inert organizations. The pathology of the neglected organization can therefore be seen as a combination of aimless motion (as in excessive dynamism) and inertia (as in excessive stability).

5.5 Summary

Robustness is a healthy characteristic for an organization. Without it, organizations would not endure. Nevertheless, pathological forms of robustness exist as well. Robustness becomes pathological when sensemaking processes become so closed that they result in inertia. Alternatively, aimless motion may arise when sensemaking processes fail to produce workable constructs.

Inertia involves a lack of dynamism. The organization becomes closed, and it no longer reacts to external stimuli. This is the stereotypical view that most people have with regard to unchangeability: an organization ceases to react to change and becomes involved only with itself.

Aimless motion is characterized by a lack of stability. For some reason, an organization can no longer manage to establish and consolidate its learning and experience. Everything remains forever new, and experience fails to be utilized sufficiently.

Part II

Tenacity

This part of the book addresses the phenomena that change agents encounter when they intervene in organizations. Robust organizations react to change with tenacity. In the first chapter of Part II, we develop a theory of tenacity; in the subsequent chapters, we describe the behavior of people within robust organizations when confronted with change initiatives. People tend to react by reverting to the old situation, by smothering the change or by acting in a calculated and self-interested manner.

6 The Tenacity of Organizations

The characteristic of robustness tends to remain "invisible" as long as organizations are in a stable situation. Robustness is only really observable in situations of change. The change agent experiences it as "tenacious" reactions on the part of the organization.

While robustness is a capacity and a characteristic of systems, tenacity refers to the system's reaction to change. In Chapter 1, we outlined the capacity of sensemaking processes both to produce changes and to strengthen the constructs that have been created. Change interventions call upon the first capacity. Change requires organizations and the people in them to start regarding the world around them in a different way, and to alter their behavior accordingly.

6.1   Tenacity

Robustness is created through the actions of people, and these actions, in their turn, are formed by the robust context of an organization. They are bound in a symbiotic relationship. To the parties involved, reality and actions appear to coincide with each other in a logical and natural way. Change interventions thus always call into question that which appears logical and natural to those involved, demanding something new or different instead. At that point, robustness, which is largely anchored in routines and behavioral patterns, no longer coincides with the actions of people. That which once held together naturally in the undisturbed situation is pushed apart as it were by a change initiative.

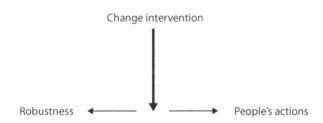

Figure 6.1   Dividing forces.

This kind of "pushed-apart" environment, in which habits and fixed behavioral patterns are still rooted in the old context and therefore fail to connect with the new context, makes people feel uneasy and insecure. They must once again find their way through their actions, creating a new robustness as they go along.

A housing corporation decided to change its strategy and concentrate more on customer satisfaction than on technical excellence. This change also involved a shift in the balance of power within the organization. Whereas the inspectors had previously had the most influence on the maintenance processes, the balance now shifted to the Customer Contacts department. Instead of being received by the inspectors, all requests and complaints were now received by that department, which would then make a decision about the requested maintenance. The aim was to gain more insight into customer demands and to enable the organization to adjust its strategy according to customer wishes and customer satisfaction. It soon became clear that the inspectors' on-site consultations with the customers had a greater influence on what maintenance work was actually carried out than did the instructions that were issued by the Customer Contacts department. According to the inspectors, the on-site consultations constituted a far more effective routine than did the new "roundabout route."

After having been pushed apart, the robust qualities of an organization and the actions of people tend to revert to the previous situation, resisting or ignoring the change intervention. We identify this reaction as tenacity, a phenomenon that is experienced primarily by change agents. As experienced by the change agent, tenacity arises because robust systems and the people in them guard that which has been constructed against change.

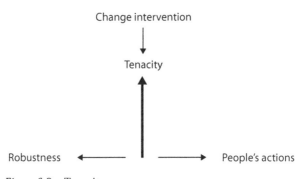

Figure 6.2   Tenacity.

6.2   Robustness and Tenacity

When we speak of an organization, we are referring to a whole, but it need not be a coherent, rationally organized whole. This entirety consists of people, islands of meaning, stories, behavioral patterns, formal and informal routines, history, memory, and other motivations that drive people's behavior. All of these factors are to some extent linked to each other and to people, and people act more or less rationally based on these factors.

The concepts of "loose couplings" and "ambiguity" as used by Karl Weick (1979) are important aids to understanding this. Weick argues that organizations are, to some extent, loosely coupled systems, in which the cohesion between components is relatively loose and can shift along with the meaning that people attach to them. The fact that the couplings are somewhat flexible makes the system elastic and therefore tenacious. "Ambiguity" refers to the fact that meaning is not fixed but changeable, and that it can refer to different contents at different times and in different places. Although the quality of robustness makes it possible to construct a workable cohesion, the exact way in which this works and the reason why it works are not always clear.

Change interventions impinge on these complex yet loosely coupled, coherent constructs, but they are unable to steer them. Their complexity and ambiguity, along with the fact that relations between components in an organization are not logically structured, makes it difficult to adapt a change strategy to them. If we regard an organization that is involved in a process of change as a body that is managed, and if we regard the change agent (a role that can be performed by either a person or an organization) as the managing system (adapted from De Leeuw; Otto and De Leeuw, 1994), the managed body that is undergoing change will usually be more complex than the managing system. This produces an asymmetric form of interaction between the change intervention and the often far more complex and layered robustness of the organization. This creates a dilemma for the change agent, who can choose to connect to either the relative simplicity of the managing system or the complexity of the managed body.

6.3   Human Shortcomings and Tenacity

Whether people want to go along with a change or not, they will be faced with a conflict between the old, familiar behavioral patterns and the new behavior that is required of them. New behavior is not always easily learned and it is always easy to fall back upon familiar behavior.

People are not ideal changing machines that are able to help shape every change or to develop their competencies to a level of excellence that allows them to realize every change. There are limits to the capacity of people to learn new behavior and unlearn old behavior, whether as individuals or in groups.

6.3.1   *Individual Limitations*
Schein (2006) argues that *"people who are confronted with the necessity of unlearning that which is familiar and learning something new, will offer resistance in order to protect their position, their identity, and their group membership, even at the price of survival anxiety and feelings of guilt."* Schein goes on to argue that a fear of the new is rooted in learning anxiety, which is a combination of several specific types of anxiety, which rear their heads each time we think about unlearning something familiar and learning something new. According to Schein, this involves the following fears:

- fear of temporary incompetence
- fear of punishment for incompetence
- fear of loss of personal identity
- fear of loss of group membership.

Learning anxiety interferes with the attempt to learn new behavior. When change interventions drive a wedge between an organization's familiar robustness and new behavior, people are confronted with these potential fears. The resistance evoked by an intervention contributes to the preservation of existing habits. Moreover, learning anxiety interferes with learning new behavioral patterns. Prompted by feelings of anxiety, the old patterns remain dominant or even become stronger than they were before.

It is not just feelings of anxiety that prevent people from learning other behavioral patterns instantly. Individuals are loosely coupled systems as well. It is not always possible to trace a direct causal relation between people's intentions and their behavior, or between what they *say* they do and what they do in reality. In our chapter on robustness, we described how constructs can take the shape of a theory of action (Argyris, 1991). A theory of action consists of a set of rules that individuals use to shape their own behavior and to interpret the behavior of others. What makes such a theory of action complicated is the fact that the things people say about their actions differs from what they actually do, often without people being aware of it. Argyris distinguishes between the "espoused theory" and the "theory-in-use."

> In many organizations, we see that people feel a need for increased openness and directness in their dealings with each other: "It would be better to talk *with* each other than *about* each other, because the way we communicate now is far too indirect. We really need to take a more direct attitude among ourselves." When we look for the reasons behind this lack of directness, we see that many people find it difficult to be openly critical of their colleagues. They are afraid this will lead to arguments and that their colleagues will no longer like them. The need people feel for more direct communication is the espoused theory. The more hidden need to be liked by others is the theory in use, which often dominates people's actual actions.

Particularly in insecure situations, the difference between people's words and actions is great. The loose couplings between intentions and behavior, between what people would like to do and what they actually do, offer the possibility of choosing the security offered by the existing order. Theories of action work in such a way that the people involved do not notice the difference between their words and actions. For example, we could wonder whether the convictions held by the politician in Chapter 1 regarding work and rest times are in line with the hours that he works himself. People fail to notice that their actions are inconsistent with their convictions regarding the same. As a result, changes take root in people's words, but not in their actions.

6.3.2 Statistical Limitations

Many changes are inspired by the ambition to make an organization operate more effectively as a whole. This involves improved cooperation, a better utilization of the available talents and skills and a more effective organization of business processes. It is not necessary for everyone to be able to do everything. At the same time, however, such a process usually demands that people be able to do more than they did before. This is where we encounter an obstacle on a statistical level.

As a collective, people are also limited in their efforts to learn, and these limitations are the reason that changes do not always go the way that we would like them to go. The ambition to improve an organization is often too large in scale. The will to change people's behavior, their competencies, or both often forms a part of that ambition. These lofty ambitions are thwarted, however, by human shortcomings. In contrast to some individuals, the average person cannot be brought to excellent achievements.

Human capacities are highly variable. We have a general idea of their distribution: most people have an average capacity for learning. There are relatively few people who are able to develop a specific quality to a very high level.

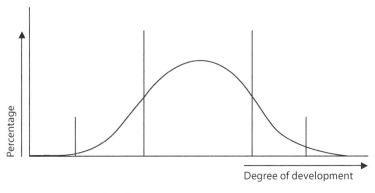

Figure 6.3   Normal distribution.

In other words, in any given organization, the majority of people will have an average command of a particular skill or quality and an average capacity for developing it. These are the materials we have to work with. High-fliers are rare. This has consequences for the way changes play out in reality. If a change is implemented by developing, improving, and reinforcing individual qualities, we cannot expect the performance of the organization as a result of the change to be higher than the average performance of its staff.

6.4   Tenacity and the Change Agent

Tenacity is not a pleasant thing to encounter when all of our efforts and ambitions are aimed at implementing change. In their turn, agents of change react to the tenacious reactions described previously. It is not always an option to resign ourselves to the reactions we encounter, or to take our time to perform a thorough analysis of the phenomenon and adjust our approach accordingly. The change needs to be carried through. When we try to step on the pedal and carry the process through at the same pace as planned, the effect is often increased counter-pressure. The change is increasingly seen as belonging to the change agent, while the old situation is increasingly identified with those who are required to change. The tenacity manifested in the interaction between the two parties continues to grow.

6.5   Reaction Patterns

Tenacity is never the result of either people's actions or of robustness alone. The way in which these aspects are held together produces a coherent reaction pattern in which the actions of people are interwoven with the robust aspects of an organization:

- The robustness of the organizational aspects that are the targets of change and the actions of the people involved is such that the change has a temporary effect but eventually springs back again.
- People bring the change in line with what they consider to be realistic in order to ensure that robustness is retained.
- People are unable to connect fully to the change and, aided by their knowledge of the organization, they display calculated behavior in order to retain existing patterns.

6.6   Summary

For change agents, tenacity is the most recognizable and most tangible form of unchangeability. It is an attitude that organizations can adopt in response to changes that threaten their robustness. Tenacity arises in situations in which the approach to change is not sufficiently rooted in the capacities and limitations of the organization and its people. It is often the product of a confrontation between a relatively simple model of change and a complex and layered reality.

The reactions of organizations to change initiatives tend to fall into a number of tenacious reaction patterns, to which change agents react in their turn. Tenacity can pose a significant problem for change agents, as their reaction often aggravates tenacity instead of reducing it.

7

Springing Back

Not all changes take root. There are situations in which a change is set in motion but later springs back to the way it was before. This kind of flexibility can also be observed in nature. We therefore asked ecologist Louise Vet whether she recognized this phenomena and whether she could provide us with some examples from her field.

Change for the Sake of Not Having to Change: A Metaphor from the Field of Ecology

Prof. Louise Vet

Professor in Evolutionary Ecology at Wageningen University and director of the Netherlands Institute of Ecology (NIOO-KNAW)

The characteristics of all livings beings on earth are largely established genetically in their DNA, which forms an elegant and unique "barcode" that distinguishes species and all individuals from each other. Genetically, each individual is slightly different from the next. This variation forms the playing field for the process that we call natural selection. In a favorable environment, most individuals have a considerable capacity for reproduction. A fly can lay around a thousand eggs during its lifetime. If all of these eggs were to reach maturity, however, the whole planet would be covered in flies within no time. The reason this does not happen is that, in every environment, the majority of individuals die before they are able to reproduce. Some of the "survivors" are somewhat more successful than others are. The winners have the most descendants and pass on their genetic makeup to future generations; we know this process as "survival of the fittest." The process of natural selection has led to the fantastic adaptations in behavior, form, and function that we see everywhere in nature: the incredible speed of the cheetah, the streamlined shape of a fish, and the sharp eyesight of a hawk. These traits are genetically recorded adaptations to the specific environments in which the organisms live and function. Polar bears on the polar ice, bacteria that live at 130° C around volcanoes in the depths of the ocean, and drought-resistant cacti in the desert: all of these are the result of evolution through natural selection.

A strong genetic adaptation to a specific environment has pronounced advantages for an organism if this environment is stable: only then will a genotype shape an organism with the greatest chance of survival and reproduction. If the environment changes drastically and suddenly, however, things can go badly wrong. In the case of severe disruptions such as the destruction of habitats, which are now usually the result of human interference, there is no escape. The usually slow process of evolution through natural selection cannot provide an answer to such changes, and the species is bound to disappear from the area or, in the worst case, die out entirely. After all, a fish on dry land has no chance at all, and if the tropical rainforests disappear, orangutans will no longer be able to survive in the wild. In his documentary *An Inconvenient Truth*, Al Gore points out the negative effects of climate change on biological systems. He illustrates these observations with Dutch research on songbirds (Netherlands Institute of Ecology, NIOO-KNAW). Now that spring is starting so much earlier in the year, the birds miss the highly temporary peak of caterpillars, which now emerge earlier as a result of the higher temperatures. These caterpillars are the single most important source of food for young songbirds.

Even though there is genetic variation between individual birds with regard to the date on which they lay their first eggs, the range of variation in egg-laying dates among birds is not sufficient to adapt to this rapid climatic change. The result is a drastic reduction of the breeding success of the birds.

Not all environmental changes have such drastic effects. Nature is subject to variation, and small environmental changes are common, also during the lifetime of an individual. One year can be warmer, wetter, or richer in sources of food than the next. Organisms manage to deal with these fluctuations through a process called "phenotypic plasticity": a capacity for limited adjustment during the life of an individual. A houseplant that is put in a dark spot will grow longer stalks in search of more light. If we stay in the mountains for some time, we start to produce more red blood cells, making it possible to maintain our oxygen levels. When we perform heavy labor, we develop calluses on our hands. These adaptations are not genetic, and they are not transferred from parent to child (with some exceptions called epigenetic traits). An individual's *degree* of plasticity, however, *is* a matter of genetics, and it is the result of evolved adaptation to a naturally variable environment. Some individuals have greater plasticity than others do, and this can have important consequences for their "fitness."

Although this seems to be a good system, such adaptations to what is frequently a suboptimal environment do not come without a cost. The energy that an organism must invest in change cannot be invested in other essential matters. It is a case of bending in order not to break. If the pressure exerted by the environment diminishes and the need for phenotypic plasticity decreases, the organism will quickly revert to the old, genetically determined pattern.

The searching behavior of parasitic wasps, so-called parasitoids, provides a good example of temporary adjustment and a consequent return to innate behaviors. These parasitic insects lay their eggs in other insects, which are euphemistically termed "hosts." Once a host has been parasitized its days are numbered, and it will therefore do anything it can to prevent being found by the wasps. But parasitoids are clever and use the odor of the preferred diet of their hosts in order to find their victims. The wasps have an innate preference for the odor of the food (for example, a particular type of plant) that their hosts usually feed on. If these plants are not available because of temporary changes in the environment, the female parasitoid is faced with a problem. She will have to search for alternative plants that her host might also feed from. If she manages to search these plants, she will be rewarded with a successful parasitization, which will produce offspring. As she lays her eggs, she learns the odor of the new plant and, in some cases, develops a preference for that odor over her innate odor of choice! This form of learning by association resembles the well-known experiments by Pavlov, who taught his dog to associate the stimulus of the sound of a bell with food (the reward). In its next search, the wasp will be able to use the learned odor if more of these plants with hosts are present in the environment. If this is not the case and consecutive searches do not produce rewards, the wasp will forget the newly learned plant odor and revert to her inborn odor preferences. They are just like people . . . .

7 Springing Back

This chapter is about situations in which changes appear to have been carried through successfully, but which ultimately spring back to the way things were before. It is about changes that are set in motion, effectively translated into what they actually mean for the people involved, and seem to have support, but nonetheless revert to the old situation.

Louise Vet describes the way in which ichneumon wasps learn to recognize the scent of plants other than the ones with which their genes have made them familiar. The wasp is capable of learning—for as long as it is necessary. If the "second–choice" plant ceases to be available, the acquired learning will soon recede into the background, and the wasp's behavior will spring back to its old preferred behavioral pattern.

We also observe these kinds of patterns in organizations. Just as the ichneumon wasp learns new behavior only to revert to its own, genetic predisposition, organizations also fall back into their original behavioral patterns. The example of the ichneumon wasp does not illustrate a failure to learn, but a reversion to familiar patterns. The example of the birds who can no longer find food because the caterpillars have started emerging earlier in the year illustrates the way in which processes of adjustment are delayed because certain patterns continue to invoke old behaviors. This phenomenon can also be observed in organizations.

In an e-mail we once received, an ex-colleague described the way in which the situation in her own organization had sprung back to its old patterns: *"Here in Amsterdam there are plenty of developments. This organization wouldn't be itself if it wasn't slowly sliding back into the old patterns that existed before you came (with the board taking over the duties of management, management taking over the duties of the HR department, the HR department preferring to playing 'duck, duck, goose,' and the head of Unit 7 is out of control). . . . In other words, there are still many laughable situations, along with many things that are no laughing matter."*

7.1 Robustness and Change

7.1.1 A Fading Capacity for Change

In processes of change, insecurities are particularly likely to arise. Old and trusted routines are thrown overboard, and the new behavior is not yet familiar. The old meaning loses its value and there is a need for new certainties or, in other words, new meaning.

In processes of change based on a vision of the future moving toward a beckoning perspective, charismatic leaders and enthusiastic front runners usually play an important role. They can invite people to go along with the movement that is desired. If the members of an organization have faith in these agents of change, they will usually be given the benefit of the doubt and the space that is needed to set the change in motion. If they are successful, they can serve as an anchor for people; they can help create new meaning, demonstrate new behaviors, and help them to accept the new behavior and render it permanent. Gioia and Chittipedi (1991) describe an iterative process of "sense-giving" and "sensemaking" in organizations. Sense-giving is what change agents or management figures do to support the process of changing meaning. Charismatic leaders reduce insecurities by offering attractive new meaning. New meanings are provided as alternatives to the meaning that was already present, and they operate as a guiding model for new or adapted meaning. This presents a source of competition for the old meaning. People can weigh both options in their own sensemaking processes. If trust in a change agent can initiate a process of sense-giving and sensemaking, the change will be set in motion. People will change their "superficial" meaning, but they will still require the help, vision, or presence of a powerful sense-giver in order to do so.

When these leaders and front-runners are no longer present or available, when the project ends, the interim manager disappears, or the supervising manager finds another job and there is no comparable figure to take their place, the result "disintegrates," and the situation reverts to the way it was. As long as there is a leader or front-runner to guide the organization in adopting the new behavior, it is relatively easy to keep practicing new behavior, thus retaining it. Sense-giving remains important, albeit to a decreasing degree, as new issues arise that require sense-giving and the exploration of new behavior. If new meaning cannot be found quickly, and if there is nobody to support this process, the old behavioral patterns will present themselves as an appealing alternative.

> In one project, the interim manager paid a great deal of attention to the way in which members of the management team (MT) interacted with each other. In the initial situation, many matters were arranged informally, there were feelings of mistrust within the team, and the members had little respect for each other. As a result, very little team cooperation took place and the organization was not managed explicitly enough. Moreover, the MT did not enjoy much esteem within the organization. In the interim project, considerable time was spent on team cooperation, on continuing to communicate with each other, and on learning to understand each other's standpoints and open them for discussion. This resulted in a more open style of communication, and it eventually led to the MT meetings becoming the main platform for discussing and coordinating important matters. All of the team members indicated that they experienced this as a clear improvement over the previous situation. In the last week of the interim manager's visit, however, when the MT had planned to devote a final conference to discussing cooperation, "it was decided" the day before that meeting not to go through with it after all, because "well, you're leaving anyway. . . ." The urge to change disappeared; old behavior (arranging matters informally) resurfaced to take its place. The counterforce had not yet gathered enough strength to resist the old habits.

WHY ORGANIZATIONAL CHANGE FAILS

7.1.2 *Old habits Die Hard*

One characteristic of problem-related changes is that people experience or recognize a problem; it is part of their meaning structure, and it is therefore in their own best interest for a change to take place. New meanings are attractive because they offer an alternative to the existing, problematic situation. They are relatively easy to create and translate into new behavior. If this new behavior actually succeeds in addressing the problem, it has the potential to be repeated and become routine. It does require consistent care and attention, though. People have a tendency to see routines less as routines and more as the logical way of doing things, and to see new behavior as unnatural and artificial at first.

When problems occur, people have problem-solving routines that they know will work. They are logical. The possibility that they might no longer be logical in a changing situation is not always self-evident. Many problems require a rapid response. In these cases, there is usually no time for reflection or consistent attention. At this speed, people instinctually switch back to their old problem-solving routines. As effective as they may have proved for solving problems in the past, and as easy as they are to fall back on, these tried and true routines keep new behavior at a distance.

> A government agency was working on a new management philosophy that was intended to create a more independent relationship with the central government. In the project, considerable attention was paid to translating the vision and strategy into new behavior and new forms of cooperation. When the financial situation suddenly took an unexpected turn for the worse, however, the government announced budget cuts. This gave people a reason to stop paying attention to the change that was being implemented. Even though the organization's funding was entirely independent of the aforementioned change and need not have had any negative effects, the process still ground to a halt. People fell back into their old management patterns. The most important reason was that they knew how to use these old patterns to "arm themselves" against the cuts in a clever and convenient way. These routines were more attractive than was the option of trying to approach the cuts in terms of a new management model.

7.2 Human Behavior and Change

People can also encounter situations in which the patterns and routines of their environments are more forceful than their own will to change. The robustness of the organization is greater than the sum of the people who work within it. It has a longer history, and it has been worn into all aspects of the system. Even if individuals have changed their behavior, the pattern of the organization has not necessarily been adjusted along with these changes. As much as people in the organization may want the change, more is required to anchor the new behavior in the organization's robustness.

7.2.1 *Not Invented Here*

Learning and development are hot topics. People are regularly sent on training courses or offered the services of a coach. Coaching is currently so prevalent that people can be divided into two categories: those who coach and those who are coached. The Netherlands is peppered with conference venues where people put a great deal of effort into learning new things. We attend congresses and conferences, we go to courses and workshops, we are coached, and we network endlessly. Participants return to the workplace with fresh insights and intentions to do things differently. In the workplace, however, their behavior is not always recognized. Many people have told us about the disappointing reactions that they have received from the people who stayed behind, "You're acting kind of funny; have you taken a course or something?" The new behavior meets with incomprehension on the part of the people who were not there. After all, they were not privy to what happened "on retreat."

There is a strong tendency for individuals to spring back to their old behavior because their environment demands it of them. After all, one employee's new behavior demands a new response from colleagues, but they cannot see why a new response is needed or how they are supposed to respond. What they actually ask is for their colleague to revert to normal, to the way things used to be. In the end, the employee is resigned to doing just that, as it is of no use to lance at windmills.

Even if the behavior has been requested by the organization or by the people themselves, behavior that is truly different is experienced as a challenge.

> The following happened to a middle manager who was being coached. The coaching was aimed at making him take a more assertive stand with regard to his staff, develop more of an individual course and direction of his own, and take others along with him in that direction. None of these behaviors was among his strongest points. While working on the convictions underlying that behavior, the manager in question came to a number of insights into his role, his behavior, and his context. He adjusted his behavior accordingly, becoming more assertive, actively approaching people, and making decisions. This constituted quite a change for the staff involved. Even though he had included them in his developmental process and had informed them of his learning goals and the types of changes they could expect to see, one staff member felt particularly threatened by this new behavior. She went to the director to complain. The director responded by going to the middle manager and telling him that, on top of his previous demands, it would be better if he acted more like a people manager and that he should avoid making his people angry. The middle manager responded by reminding the director of his own wishes regarding his skills, the choices he had made, and the way his behavior reflected his learning goals, but his efforts were in vain. In a consultation involving the director, the middle manager, and the coach, the coach explained that people in the manager's environment needed time to become accustomed to the new behavior, and that their unfamiliarity could produce these kinds of reactions. Once again, the director failed to

respond to the arguments. In the end, the middle manager simply gave up on his new behavior.

7.2.2 Back to Square One

If an individual change is set into motion, or if a small group of people develops new working methods, it does not mean that the change has been anchored within the robust fabric of the organization. A change in the balance of power can completely undo such a budding change.

Consider the example of a conflict between the board, management, and works council of a Chamber of Commerce, which produced unworkable relations, as all parties regularly encroached on the domains of the others. One attempt to resolve the conflict involved working on clarifying the division of roles and the corresponding cooperation and interaction between parties. All those involved were happy that the other two parties were also held partly responsible, and all three tried to be more aware of their positions and contributions. The process gradually brought about an improvement in the situation. While that process was in full swing, the periodical elections for the board came around. The old board, which had come to see itself as just one of the parties and felt jointly responsible for solving the existing problems, was replaced by a new board, which felt it had nothing to account for and reverted to the old ways. The existing power relations and the game that went with it suited their position. The fact that the board reverted to the old pattern caused the other parties to do the same. The old pattern organized the balance between the parties, and that balance needed to be restored . . . to the old situation.

7.3 Summary

Springing back is one of the manifestations of tenacity. It is a delayed form of tenacity, as the organization does change at first. Springing back occurs in situations in which there is a great willingness to change, although the available force for change is limited in comparison. As soon as the attention paid to change (i.e., the "external" force for change) wanes, the situation springs back to the way it was before. Springing back has to do with the fact that new meaning and new patterns of behavior have not yet sufficiently taken the place of the old ones. They have not yet taken root, and the old patterns are still anchored in the subconscious actions of the people involved.

8

Smothering Change

Appeals for change usually come from a different part of organizations than the place where changes have to be carried out. Changes also do not always correspond to that which people value or find important, or to that which people are capable of changing. We are familiar with Hans Bennis's ideas about the limited influence of language managers and the degree to which the language develops autonomously, which is why we asked him to write about the changeability and unchangeability of the Dutch language.

On the (Un)changeability of Language

Prof. Hans Bennis

Professor of Dutch Language Variation at the University of Amsterdam and director of the Meertens Institute

What is the correct formulation: *"U hebt betaald"* ("You[polite] *have*[second-person singular] paid") or *"U heeft betaald"* ("You[polite] *has*[third-person singular] paid")? Most speakers of Dutch think that it is possible to give an answer to this question. They do not know the answer themselves, but they presume that one of these sentences must be the correct one. However, it is not as easy as that. The most authoritative grammar of Dutch, the Algemene Nederlandse Spraakkunst, or ANS, fails to give a decisive answer in favor of one of these variants. ANS takes both variants to be acceptable. Other grammars show similar indeterminacy.

Here we encounter a common misunderstanding about natural language. People often think that language is a fixed system of agreements that have been decided on by language managers in order to govern correct communication. In this view, grammar is taken to be a culturally determined system of rules that we struggle to master at primary school. The correct application of these rules would make it possible to decide in cases such as the "U hebt" vs "U heeft" issue just described.

From a scientific perspective, this point of view is untenable. Language is part of a cognitive organization that is developing continuously without conscious interference. The grammar of a language is something a child acquires spontaneously in the first years of his life. It allows the child to speak its mother tongue. Looking at language from this perspective, acquiring a language is a natural process that is governed by cognitive principles. We might want to be able to influence the process of language acquisition in order to ban sentences such as *"Hun hebben gewonnen"* ("Them have won"), instead of *"Zij hebben gewonnen"* ("They have won"), but that appears to be wishful thinking.

"Hun hebben gewonnen" is generally considered to be a bad, ugly, sloppy Dutch sentence that violates the grammar of the language. The frequently occurring use of the object pronoun *hun* in subject position is the cause of much aggression and frustration. Politicians such as the former minister of education Plasterk promise to take measures against the use of the "wrong" form of the personal pronoun. However, it remains to be seen whether this struggle against bad language will have much effect. This anger against the deterioration of the standard language is something of all ages. In his 1865 essay collection *Klaasje Zevenster,* the famous nineteenth-century writer Jacob van Lennep wrote, "You [polite] in the nominative Case is a sin against Dutch Grammar". He was fighting against "awful, ungrammatical sentences" such as *"U zondigt tegen de regels van het Nederlands"* ("You [polite] sin against the rules of Dutch grammar"). At that point in time, there was no explicit grammar that excluded the use of *U* in subject position, but similar to the case with *hun* in subject

position, the average, well-schooled language user had preconceived opinions about good and bad language. Notwithstanding his eloquent attacks on *U* as a subject, Van Lennep has long since lost the battle. Today, nobody disputes the correctness of the use of *U* as a subject pronoun any more.

An interesting case of language change concerns the use of the third-person plural object pronouns *hen* and *hun*. In 1625 the artificial distinction between the dative pronoun *hun* and the accusative pronoun *hen* was introduced by the grammarian Christiaen van Heule. Following the example of Latin, the dominant language of that period, grammarians thought that a case distinction between accusative and dative should be present in any self-respecting language. His proposal to deliberately change the grammar of Dutch in this direction was quickly adopted by important writers of this period, such as Vondel and Hooft (Van der Sijs, 2004). Now, almost 400 years later, still nobody knows intuitively when to use *hun* or *hen*. People make mistakes all the time. The distinction remains completely unnatural unto this day. We can therefore conclude that language change that has been consciously devised by language managers has only a slight chance of succeeding. And to the extent that such changes succeed at all, they will often remain unnatural additions to the language system. We thus observe that conscious interference with the grammatical system of a language does not really help to prevent the language from changing; moreover, we saw that the conscious introduction of improvements of the language has a restricted chance of success.

Contrary to such conscious involvement with the language system, we may observe that natural language changes do occur frequently. Often, these changes are hardly noticed by the speakers of the language. A relevant example is the recent change in the paradigm of modal auxiliary verbs. The forms for second person singular are assimilating to the forms for first and third person. It used to be "*ik kan*" ("I can"), "*jij kunt*" ("you can"), and "*hij kan*" ("he can"), but within a short time the form "*kan*" has become generally acceptable for second person singular ("*jij kan*"). The same applies to "*jij zult*" → "*jij zal*" ("you will") and other modal auxiliary verbs. There is hardly any opposition against this change. It has crept into the language, unnoticed by the puritans and other admirers of correct language use. It can be observed in literary novels, quality journals and the speech of language professionals. No one has introduced or supported this change. No language manager has been involved, either to stop language deterioration or to promote language improvement.

It is interesting to observe that the perfect auxiliary verb *hebben* "to have" appears to take part in the development that was introduced by the modals. Increasingly, we encounter the forms *jij heb* (instead of *jij hebt* "you have") and *hij heb* (instead of *hij heeft* "he has") in substandard spoken Dutch. From a linguistic perspective, this change makes sense. It reduces redundant irregularity in the paradigm of auxiliary verbs. The feature (person) disappears from the paradigm. In English, much of the verbal inflection in auxiliary verbs has disappeared already: just witness auxiliary verbs such as *can*, *must*, and *have*. However, the question whether the change is an improvement grammatically does not determine its eventual success. There is growing dissatisfaction with the reduction of inflection of the verb *hebben*. It is clearly not the case that *hij heb* can be used in written standard language or in official use of spoken language. A newsreader on the Dutch television might be fired if he used *hij heb* regularly, whereas probably nobody would notice it if he/she said *jij kan*. Why *hij heb* is considered to be a strong violation of the rules of the grammar of Dutch, while a similar process in the case of modal auxiliary verbs is generally taken to be acceptable, is difficult to explain from a linguistic perspective. In both cases it concerns an increase in the efficiency of the grammatical system.

Another somewhat related case of present-day language change is the inflection of adjectives in attributive position—that is, adjectives that are part of a noun phrase. Standard Dutch has the remarkable property that attributive adjectives are inflected with an–*e* affix, unless the noun is neuter and the noun phrase is indefinite and singular. It is *het mooiE huis* ("the beautiful house"), *de mooiE huizen* ("the beautiful houses"), but *een mooi huis* ("a beautiful house"); if the neuter noun *huis* is replaced by the synonymous non-neuter noun *woning*, then the attributive adjective shows–*e* inflection in all instances (cf. *een mooiE woning* "a beautiful house" etc.). The adjectival inflectional system is an awkward remnant of an older, richer inflectional system. Children have problems mastering the distinction neuter–non-neuter for nouns, but they succeed in getting the system right eventually. The major problem in this case concerns the export of our language. It turns out that the inflection of adjectives and the neuter–non-neuter distinction of nouns is extremely complicated for people trying to learn Dutch as a second language. In the context of international visibility and immigration we want foreigners to learn our language. In this situation it would be a wise decision to drop the completely redundant adjectival inflection and the meaningless neuter–non-neuter distinction on nouns (e.g., girl [*meisje*], woman [*wijf*], and child [*kind*] are neuter, whereas table [*tafel*], chair [*stoel*], and room [*kamer*] are non-neuter). It would seriously enhance the possibility for people to learn Dutch at a later age. However, at present mistakes in adjectival inflection and the choice of the wrong determiner (*het* for definite neuter and *de* for definite non-neuter) are considered to be strong indications that the speaker has not acquired the language adequately. From a linguistic and socio-cultural perspective it makes no sense at all to preserve these redundant distinctions, but it will probably take a long time before *een mooie huis* or *de huis* will be considered to be acceptable Dutch.

From the preceding we might conclude that our society militates against language change in so far as these changes are salient enough to be noticed, and that conscious change of the language is generally unsuccessful, but that the natural process of change cannot be prevented. However, conscious interference of language managers is not always doomed to failure. A striking case concerns the reintroduction of the reflexive pronoun in the sixteenth and seventeenth centuries. In the Middle Ages most Dutch dialects no longer had a reflexive pronoun. One finds sentences such as "*Hi wast hem*" ("He washes him/himself") whether or not the agent *hi* is washing himself or someone else.

Around 1600 the reflexive pronoun was introduced quickly. The language area fell apart into three smaller areas. The eastern part of the area took over the High German *sich(selbst)* and turned it into *zich(zelf)*. Frisia and Flanders opted for the English-type reflexive *hemzelf*, whereas the western and middle varieties chose a new variant, *zijn eigen* (lit. "his own"). Recent research shows that the dialects still show these different variants (cf. Barbiers et al., 2005).

After the Fall of Antwerp in 1585, Holland (i.e., the provinces of North- and South-Holland) was the dominant area in the Netherlands in political and cultural respect. Hollandic language varieties thus heavily influenced the development of a Dutch standard language in the early seventeenth century. The language managers that were given the job of developing a standard language—partly for reasons of writing a bible that could become the standard in the whole language area, the so-called Statenbijbel—chose the eastern reflexive pronoun *zich(zelf)* over the Hollandic/Brabantish variant *zijn eigen*, for reasons unknown. This choice proved a success, as it is completely undisputed for standard Dutch. In fact, using the Hollandic reflexive *zijn eigen* in the standard language is considered to be bad form these days.

So it appears to be the case that there are possibilities for conscious involvement in the organization of the language system. Trying to stop natural developments is doomed to fail, as is introducing changes that are new to the system. However, if a choice is consciously made between available natural variants, the language manager stands a much better chance of succeeding. This indicates that a modest attitude on the part of the manager will produce the most successful results. In how far this is true for managers of other types of organizations, is a question that I leave for the reader to find out for him or herself.

References

Barbiers, S., H. Bennis, G. de Vogelaer, M. Devos, & van der Ham, M. (2005). *Syntactic atlas of the Dutch dialects*. Vol. 1. Amsterdam: Amsterdam University Press.
Haeseryn, W., et al. (1984/1997). *Algemene Nederlandse Spraakkunst* (ANS). Groningen: Martinus Nijhoff.
Van der Sijs, N. (2004). *Taal als mensenwerk; het ontstaan van het ABN*. Den Haag: SDU.

8 Smothering Change

The most important insight we derive from Hans Bennis's contribution is that users tend to see that which already exists as good and consider changes as examples of the system's degeneration, unless a change has developed in practice—in this case, in the practice of language. Prescriptive alternatives or rules usually fail to catch on unless they latch onto something that is already in use, in which case there is a good change that they will take root. In other words, language users are the main determining factor in the language development. Language managers are relatively powerless in comparison.

Like languages, organizations are subject to change. Staff and managers are constantly making decisions to do things differently or to adopt a fresh approach. Some change initiatives originate on the shop floor. In some cases, these initiatives are adopted and implemented on a wider scale. This kind of process usually proceeds organically.

Just as in the history of languages, however, other developments that take place in organizations have been designed on paper, or dictated from above. These changes are not always in touch with the daily practice of an organization. In the worst case, they can actually make it impossible to carry out this daily practice. Within the context of these tensions, the users of the devised plan (i.e., the employees) determine how it can actually be put into practice. They evaluate it according to the degree to which the planned change is usable.

As with the previous chapter, this chapter is not concerned with examples of conscious stalling or opposition to change. The focus here is on the discrepancies that exist between organizations on paper (i.e., according to those who manage or direct the organization) and the organization at the level of the user, and the way in which people reconcile these kinds of differences. We explore the way in which changes are sometimes smothered through good intentions and impotence.

8.1 One Must Make Shift with What One Has

Changes sometimes make demands that conflict with an organization's capacity for adaptation or put pressure on its working processes. In many cases, these changes are not particularly useful or convenient to employees in relation to their "usual tasks." The organization's employees are faced with a dilemma: they can be loyal and go along with the change, which will inevitably involve many things going wrong, or they can opt for a strategy that will "smother" the change in the limited possibilities of the organization. Employees adjust changes to fit in with what is possible, as they are not able to adjust the organization to what others feel is needed.

8.1.1 A Blessing in Disguise: Constructive Disregard

Although we do not always realize it, the daily practice of organizations diverges from the way operations are described in the official protocols. These differences are "covered up." It does not help an employee or middle manager to keep insisting that the practices that are described do not work in the way that they are intended. Employees who rigidly adhere to the duties specified in their job descriptions tend to make themselves unpopular. Such attitudes are seen as resistant, and therefore as negative and disloyal. For this reason, most people simply make minor adjustments in the course of their work.

Organizations are not really designed to uphold every rule that is recorded in their procedural manuals. Rules can be contradictory or impractical, or they may have bizarre consequences. Employees are aware of this, and they try to achieve a reasonable interpretation of the rules in day-to-day practice instead of following them to the letter. Their choices are pragmatic. This really comes to light when a work-to-rule action is held. Work-to-rule is a form of industrial action that involves following all regulations to the letter. This results in the system grinding to a halt very quickly indeed. It could be said that a work-to-rule action involves employees taking the managers and their rules more seriously, or in any case more literally, than the managers wish.

The consequences of such an action demonstrate the impracticality of applying all the rules. Some change initiatives can even contain unworkable aspects. Some have not really been thought through in terms of their practical application; others place pressure on processes that are otherwise functioning successfully. If such a change were to be introduced as planned, the organization involved would grind to a halt in no time.

In order to make the reality of formal organizations, strategy papers, policy memos, and procedures workable, employees make adjustments to existing routines. The existing routine is "tweaked" a little according to the altered requirements, but not so much as to threaten the robustness of the system. In this way, the working process will continue to run smoothly, the results will remain acceptable, and the clients will stay satisfied. Choosing not to implement changes to the letter constitutes a highly constructive and sometimes brave contribution, and it is often motivated by feelings of loyalty.

8.1.2 Time and Priorities: Production Overload

Change assignments often come on top of the normal workload. In practice, therefore, the implementation of changes clashes with the need to maintain regular production. Employees must set priorities, and they often assign greater weight to daily production than they do to carrying out changes. After all, they are held accountable primarily for their daily production.

> Consider the example of a change project in an organization that was under considerable pressure from politicians and society, and in which there was considerable work pressure. Because the social and political "eyes" were trained upon them, the board was forced to try to obtain a

better grip on the organization's management and performance, especially in terms of reducing waiting lists. To this end, they looked for instruments, such as reporting systems, management training, and a quality assurance system. All of these extra tasks descended on the middle managers and staff almost simultaneously. At the same time, however, these employees were facing high staff turnover, a steep rise in the amount of work, and a great deal of overdue maintenance. It was too much. What happened in practice was that people chose to prioritize their "regular" duties. As a result, the changes proceeded slowly and with difficulty.

The management of business processes and the reward systems usually proceed according to variables relating to daily production. The management and rewarding of change and innovation is usually carried out much less effectively. This encourages people to focus on their "designated tasks." The system rewards us for performing our regular duties well, after all, but not for changing. In the preceding example, the middle managers were evaluated according to whether they had achieved their targets and had succeeded in reducing the waiting lists. They were not evaluated according to their contributions (with or without their employees) to realizing the new organization with all of its new instruments, or according to their presence at all of the workshops. Not surprisingly, the middle managers stayed away *en masse*, using the priority of the regular workload as an excuse.

The result of such situations is that the change becomes the spiritual property of the project leader or the top management, while the regular workload remains the domain of the middle management and staff.

8.1.3 *The Map Is Not the Territory*
Technological developments have made it possible to record and analyze vast amounts of information. In the past few years, the need to make organizations manageable and to enable accountability regarding results has resulted in complicated cycles of planning and control. Most organizations work with annual and multi-year plans, accompanied by annual and quarterly accountability reports. All of these plans and reports are replicated at each level of management. The assessments that managers conduct with employees are often based on these kinds of documents.

As a rule, when we set something down on paper, we significantly reduce its complexity. The daily practice of work is far more complicated than are the plans that are recorded on paper, and many more things happen than are reflected by the reporting criteria. As a result, the organization on paper and the organization in reality can come to form two separate worlds. The paper organization describes plans that are out of touch with what is actually happening; the accountability reports provide information about what the outside world wants to see, but they do not always guarantee that things are really going well in the organization. A common reaction, even in situations involving change, is for people to treat the paper reality as a paper tiger. We go through the motions because it is required of us, but we do not exert the extra effort that is required to connect these exercises to the way things actually work in the organization. This costs the least amount of time and is also the safest strategy.

8.1.4 A Myriad of Crossroads

Organizations can expect their employees to be customer-oriented, or results-oriented, or environment- or development-oriented. Others we have encountered are governance-oriented, quality-oriented, and production-oriented. Most of all, we regularly see that some combination of these orientations is placed on the shoulders of employees during a process of change. Instead of choosing one aspect upon which to focus, the change agent chooses an entire range of priorities at the same time, thereby asking too much of the employees. It is almost impossible for employees not to feel a bit like Alice in Wonderland, standing at a crossroads and trying to find the way:

> "That depends a good deal on where you want to get to," said the Cat.
> "—so long as I get somewhere," Alice added as an explanation.
> "Oh, you're sure to do that," said the Cat, "if only you walk long enough."
> (Carroll, 2007)

If the change agent fails to choose a focus, the employees will choose their own. However, their choices will be based on the routines that are most precious to them and those that they believe will ensure the greatest degree of continuity in their work and their particular ways of working.

8.1.5 Not Everyone Can Be a Jack-of-All-Trades

Changes are initiated to improve the functioning of organizations. Although the focus of improvement can be on the technical aspects of an organization, it usually affects the demands that are made on people as well. People are usually asked to contribute more than they did before the change was implemented.

Organizational theories often focus on the need for people to keep developing and remain employable in a rapidly changing world. This is appropriate, as the environment continues to make new and greater demands on people, who are required to "keep up." This approach to development can be taken too far, however, making it counterproductive. An excessive focus on development and improvement can decrease attention for the fact that development is not always possible.

As we outlined in Chapter 6, people are only human, and they are therefore subject to limitations. Within a population, most people will score about average with regard to particular qualities and characteristics. The Peter Principle, formulated by Laurence Peter (1969), demonstrates what happens when individuals achieve positions that make greater demands on them than they their capacities allow them to fulfill. People are often promoted according to their good performance in their previous positions. The promotions stop only at the point at which someone is no longer performing well. The consequence of this process is that people are promoted to levels at which they no longer perform well. The Peter Principle also applies to organizations: organizations work on the assumption that people will be able to rise above their current level of functioning and initiate changes that are actually too ambitious for the majority of people in the organization.

The most probable effect of such overestimation is that, in formal terms (i.e., on paper), the organization is excellent, full of broadly employable people, including older people who remain motivated up to the age of sixty-seven years and managers that possess top-level skills. In practice, however, people have trouble realizing these ambitions, and they change only as much as they are able to change.

8.1.6   *Brown Jackets and Blue Jackets*

In many processes of change, middle managers play a pivotal role. They are responsible for, and connected to, operational processes, but they are also asked to translate a change to the workplace. Combining these responsibilities often proves to be a tricky task.

Paul Valens (www.valens.nl) characterizes two distinct types of managers: the "blue jackets" and the "brown jackets." The two types of managers appear to play different roles in the strategy development and organizational change of companies. Both types of middle management staff can provide constructive input as long as the management is able to create the right conditions.

> *Brown jackets can be defined as employees who have occupied a middle management position for fifteen years and who will continue to do so, in the same company, for the next fifteen. They form the hard core of the company's middle management.*
>
> *Blue jackets are employees who are only temporary members of middle management; they're passing through on their way to a top job. When it comes to strategy development, brown and blue jackets are worlds apart. Brown jackets don't like uncertainty. This is what makes them so valuable in running a company efficiently. At the slightest doubt whether something is going well or not, they won't rest until this doubt has been removed and the operation is running smoothly.*
>
> *In strategy development, however, it's vital to deal with many uncertainties at the same time. A lot of different perspectives are possible, all kinds of options present themselves and decisions mustn't be taken too soon. A strategic concept needs time to ripen.*
>
> *However interesting and enjoyable strategy development may be for blue jackets, it is anathema for the regular middle management. They are used to analyzing things quickly and intervening decisively, and that's just what you don't want in strategy development. Strategic development is often an unfamiliar activity for which they feel no natural affinity. Blue jackets are interested in achieving the maximum amount of innovation. For them, every successful innovation means a chance of a better position. An important concern for blue jackets is their own personal strategy: they're on the way to the top, and their strategic problem is how to get there. Each change means an opportunity for them and is therefore an improvement. Brown jackets don't have that problem. Their strategic problem—or at least a major one—is how to retain their position.*

This is why they start by resisting strategy developments: they're afraid that a strategic change means they'll lose their place. They achieved their positions through experience in the current situation, but there's no telling how much this experience will be worth in a new strategic situation. For them personally, each change seems a change for the worse, even if in objective terms it may represent an improvement for the company.

Processes of change often lean heavily on the blue jackets. These agents of change, who often are in fact dressed in blue jackets, tend to consider the brown jackets as a bit of an obstacle. They are certainly loyal, but they also tend to complain. They are certainly not early adapters by any means. However, it is those same brown jacket types who are the first to realize that a change is not working, or that it may be becoming too complex to handle. It is also the brown jackets who understand that not all employees have the potential to become excellent and that it is necessary to "work with what you've got." They adjust changes, bending them into workable variants, because they are not as easily distracted by enticing visions of the future. They ensure that processes remain robust.

In the past few years, the career policies of organizations in the Netherlands have changed dramatically. Young, highly qualified, and talented young people with little experience are quickly promoted to relatively high management positions in organizations. Especially in large organizations, this creates a large middle-management layer of brown jackets of a certain age: good managers who understand the way the organization works, how to motivate people, and, especially, what the possibilities and impossibilities are with regard to change. These qualities stand in contrast to the young, ambitious people who are appointed to the management positions above them.

Once, in a discussion with middle managers about the phenomenon of "brown and blue jackets," the conversation turned to differences between "then" and "now." When these young, blue-jacketed men and women moved into management positions at an early age, they were appointed to positions where "they couldn't do much harm." Any grandiose plans and overblown ambitions were stifled by the context and tempered by the management layer directly above them. You were only promoted to a higher position when you were ready for it. This was also consistent with the expectation that you would stay with the same employer for years to come. This practice turned these men and women into realistic managers, oriented toward change, but with knowledge of the organization's limitations.

At present, another staff policy is in effect, and there is a relatively large group of young, highly qualified people who aspire to management positions. Being a manager is no longer something you grow into, but rather a position that is suited to you if you have the right kind of ambition and the right education. The managers with whom we spoke had seen how young blue jackets were being appointed to positions above them with relatively little experience and real knowledge of the organization, but with plenty of ambition. These young managers were also being charged to carry out some relatively drastic changes. It is important to note that all of the managers taking part in the discussion were people with a great sense of commitment, who were not bitter and, who were well disposed to the youngsters above them. There was no evidence of envy on their part. What they did show was concern, however, about the ambitious men and women who often push through their lofty plans despite repeated warnings from the brown jackets. Many a young and inexperienced manager has left a company, leaving behind considerable debris and people who have been damaged in their wake. It is the brown jackets, the stable factor within the organization unit, who are left to clear away the debris.

This does not mean the brown jackets suffer too much, however, because they can also take great satisfaction in describing the habit they have, when another green and unknowing boss is appointed, of sending them on a couple of impossible "errands" that have been attempted in the past but have never really come off the ground. You never know: now and then you might even encounter someone who can actually deliver the goods.

8.2 Summary

Many changes are smothered when they encounter the reality of their practical workability. Employees are eager to introduce the intended changes, but the change activities must yield to the realities of day-to-day practice. In addition, many change initiatives are aimed at a more effective utilization of "human resources." Based on the idea that people have many more potential strengths than they currently have the chance to show, initiatives are aimed at a growth in the possibilities of the people involved without taking enough account of the statistical reality of the average capacities of the majority.

It is particularly in the middle-management layer and on the shop floor that the practical implications of these management wishes come together, and these are therefore the places in which choices are made with regard to what is and is not going to happen. In this way, and with the best of intentions, numerous inspiring plans for change are smothered in the tenacity of reality.

9

Calculating

People are not pawns who simply play the role that is demanded of them in a change process. They are also individuals with their own opinions and interests, on the basis of which they determine their behavior. This behavior in turn contributes to the robustness of organizations. Insight into unchangeability is often a matter of hindsight. For this reason, we asked archaeologist Saskia van Dockum to look back on what has proved changeable and unchangeable in the history of the Low Countries.

Inertia and Change: Lessons from Archaeology

Saskia van Dockum
Director of The Utrecht Archives

Archaeologists construct stories about the past based on the pieces of a vast puzzle. They know that they do not have all the pieces and that it will never be possible to complete the puzzle. Sometimes meticulous research allows them to uncover a great deal at the level of the "square inch." One example is the research conducted into the twenty-five-meter-long ship that was found during construction work on the new suburb Leidsche Rijn, just outside Utrecht, in 2003. The archeologists called the ship a "treasure trove of information." Analysis of the timbers has shown quite accurately in which year the three oaks the ship is made of were felled (around AD 148); that the trees did not come from one and the same forest and that originally they were over forty meters tall. Furthermore, the discovery of the ship also produced a wealth of information about shipbuilding techniques and life on board.

Apart from the fact that the puzzle is incomplete, the bigger picture we are trying to piece together changes over the course of time. This is beautifully illustrated in *A Street Through Time*, a children's book which visualizes the history of a street in fourteen intricate pictures showing the same section of landscape at different points in time. The book shows the way in which historical developments are driven by both continuity and change. The first picture in the book shows the landscape long before the start of the Christian era, with endless forests and a small tribe of people who have put up their tents made of animal skins on the banks of a river. The tribe in the picture does not live in this spot permanently as they are nomadic. They support themselves by hunting wild animals and by gathering all manner of other foods such as seeds and berries. At this time, people and their households were highly mobile and would move around according to the seasons and the migrating herds. This must have required considerable effort, especially from our own point of view. The territory that formed a tribe's living sphere was extensive.

The second picture shows a scene in which a considerable part of the forest has been cut down to make way for small cultivated fields and meadows where sheep are grazed. In the same place on the banks of the river where the tents stood, a small village has now arisen, which is surrounded by an enclosure to keep out uninvited guests and wild animals.

From our own current-day perspective, which is dominated by the idea of progress, you might expect that such a change would have taken place quite rapidly, as we assume that people prefer to leave a physically arduous life behind them, if the natural circumstances and their knowledge allow for it. However, archeological research has revealed that when hunter-gatherers came into direct contact with agrarian communities living in permanent settlements, they only very gradually adopted the latter way of life. This is actually rather an understatement, as the transition took a whole millennium in large parts of the Netherlands! If you try to put yourself in the shoes of a hunter in the Stone Age, however, you can see that this is not an easy transition to make. A timber house provides a sense of continuity, but it also represents a radical change in itself. Instead of migrating with the seasons and following herds of wild animals, it meant staying in one location and trying to bend the seasons to your will. And what do you do when the harvest fails? Looking back, since the definitive change from a hunter-gatherer to an agricultural existence, there has never been another such fundamental revolution in the history of the human race. It would have been interesting to be a fly on the wall during some of the discussions that were held by tribes at the time.

Archaeology is a wonderful field that is dedicated to describing changes. These include changes in the use of materials (just think of the designations *Stone* Age and *Iron* Age); in pottery decoration; in the plans of houses; and in the way people dealt with deaths in the community. On the basis of these changes, we try to describe and preferably to explain people's behavior. As we do so, we need to be continually aware of the fact that we are looking at things from our own perspective in the here and now, based on what we find logical or what we would do in the circumstances, or on what related fields such as cultural anthropology teach us about people's behavior.

One of the archeological periods that appeals most to people's imaginations is the Roman era. These were also turbulent times in which the inhabitants of the Low Countries suddenly came into contact with a "real" civilization from the Mediterranean world, after a period of relatively slow and continuous development. The Romans conquered a large part of the Netherlands and occupied the whole area to the south of the Rhine. Their intention had been to add the north to their territories as well, but they retreated when this proved more difficult than anticipated. In the first century, Emperor Claudius made the decision to entrench his troops behind the well-defended Rhine zone. This was a sensible move, as the North had little to offer in an economic sense, and the little it did have to offer was acquired through trade.

The Romans remained for another three centuries and more. This equates to about fifteen generations. Tens of thousands of soldiers occupied various forts and sentry posts along the border. In the hinterland villas arose, furnished with splendid porticos, windows with glass windowpanes, floor and wall heating, and murals on the walls. According to the Mediterranean custom, Baths were installed: a newfangled feat of engineering including cold and warm baths as well as steam baths. The pots and pans from the era also show the influence of southern customs. Consider the example of the *mortarium*: a vessel with grit embedded in its surface, which was used as a kind of mortar for crushing herbs, after which oil could be added to it to make a flavorsome condiment. Those herbs, which included dill and coriander, were also a novelty. And what to think of a tasty drumstick: another novelty introduced to the region by the Romans. In the field of agriculture, meanwhile, we find evidence of a considerable increase in the scale of the lands used and the introduction of new breeding techniques that resulted in larger cattle.

In the course of the centuries, the Romans adopted a number of methods for governing the area. They relocated friendly tribes to areas that had yet to be conquered and concluded favorable treaties with them, starting with the bravest and most valiant of these, the Batavians. There were obvious advantages for both sides. The Romans were provided with the support of a group of expert fighters who could control an area unfamiliar to the Romans in their place, while the Batavians received a whole raft of privileges in return. They became official allies, which meant they were exempted from having to pay taxes, and they provided auxiliary troops that fought alongside the Roman legions on various fronts. One of the most honorable tasks they were allowed to perform was to join the personal guard of the emperor.

Nonetheless, eventually these governing methods proved less than effective. Actions on the part of the Romans on various fronts met with resistance on the part of the Batavians and, for a variety of reasons, the tribe rose in rebellion against the Romans in AD 69. Initially they had gone along with them and fully plucked the benefits from this alliance, but clearly had not accepted the new rule entirely. They quickly managed to convince the other local tribes to join them in their offensive, which was entirely against the expectations of the Romans. Initially, they were very successful. They were helped by the fact that they knew the terrain and its inhabitants like the back of their hand. In a short space of time, all the encampments along the Rhine—the symbols of enemy rule—were burnt to the ground, including the large Roman city of Nijmegen, then known as Noviomagus.

Another example from the same period reveals a more nuanced, almost tolerant way of governing. The previous example has already shown how Roman rule over the territories south of the Rhine at the beginning of the millennium proved to be less than absolute. The population was required to go along with the new rulers, but did not do so wholeheartedly, especially in the first decades, and kept their own cultural values. Evidence of this can be found in the way that the local population gave their own traditional gods a Roman flavor but allowed the world of their local gods to coexist under that veneer. A fine example of this practice is the god Magusanus Hercules, to whom various altars have been found dedicated. One the one hand, this was a true Roman god, or half-god in the case of Hercules; on the other hand, the figure has all the characteristics of the head of the Batavian pantheon, Magusanus. Magusanus Hercules was not the only contraction of two deities; Mars, the god of war, also had a local counterpart with which he assimilated to form Mars Halamarthus, and we also know of Mercurius Avernus. Roman historians call this practice of assimilation *interpretatio romana*. Incidentally, the Romans also tolerated the worship of native gods such as Nehalennia, as long as people acknowledged the Roman deities as well.

Sensible types, those Romans? The Romans did manage to create a stable economy in large parts of the Netherlands, which lasted for a period of around 200 years. After that period, many of the achievements from the first centuries of the millennium, such as the villas and baths, the murals, the floor heating systems and the world of the deities, were gradually lost. There was also a degree of continuity, however, for example in the influence and importance of certain places. In the Middle Ages, a new center of power arose around the old Roman city of Nijmegen. And the Bishopric of Utrecht also had its origins in a Roman fort. The bishopric of Utrecht would endure for 800 years. Moreover, the chicken had become a permanent fixture in the Netherlands!

9 Calculating

In many cases, planned changes are not changes that the people involved aspire to achieve. Many changes happen "because someone else thinks it's a good idea." We often hear about "the boss's pet project," "HR's latest invention," "bureaucratic red tape," and "the merger that the shareholders want." It is certainly true that not all changes are perceived as useful changes.

It is only very seldom that one encounters an organization that consciously assigns priority to settling down and allowing recent changes to take root and become permanent. Accountants appear to be the only ones who still refer to "consolidation." Many directors and managers seem unable to resist stacking one change on top of the other. This chapter explores the ways in which people cope with changes that are not "theirs" and in which they would prefer not to be involved.

Similar to the Batavians in Saskia van Dockum's story, who had not asked to be relocated (even if they did receive all manner of privileges in return), employees often do not ask for the changes that come their way. In another analogy, much like the Batavians initially went along with the changes, only to strike back in the end, employees also display calculating behavior in organizations.

Being familiar with the local robustness of your organization gives you the space to engage in the "game" of calculating behavior. On the surface, this game appears to center on change, but it actually amounts to no more than a shadow play. The objective is to change just enough to show others you are on board, but not enough to provide any meaningful support for the change. It is a sort of shadow boxing. One particular term for this game is employed with a heavy dose of cynicism in some organizations. Originally army slang, the term "BOHICA" ("bend over, here it comes again"), is used to warn each other that another threat is coming their way, for which they should take cover.

9.1 The Calculating Employee

We chose the title for this chapter because we want to show that people are shrewd operators who look out for their own interests. They are calculating employees. They are not mere pawns in the management game. Neither are they capital that can be invested in the same way as equipment and resources. People are aware of the leeway that they have, and they know how the game of organization works. This helps them to realize their personal ambitions and to develop their own meaning and viewpoints.

This situation implies that people go along with changes wholeheartedly in some cases, while they do not in others. They know when they "have to" bend and where there is room to maneuver. They know how to behave in order to give change agents the impression that they are going along, without actually doing so. In this respect, they are consciously shadowboxing in order to throw others off their guard and to create elbowroom for themselves. They change in order to avoid being changed.

In an interview in a Dutch daily newspaper, Friso de Zeeuw explains how people in the GDR used this kind of strategy (Schreuder, 2008):

> I discovered that, alongside the official communist system, a fully functioning informal economy existed, which was run by the people themselves. Another thing that struck me was the selective forms of shopping that were available under the regime. People were provided with employment, childcare, education, healthcare, and affordable social housing. People went along with the regime, but certainly no more than necessary. This meant attending the meetings of the factory committee, but not actually joining in the discussions. They stayed close to the edge ... The GDR regime was completely focused on structure and supervision. The country's citizens, meanwhile, looked for places that offered some leeway. They shaped this into their own repertoire, their arsenal of interventions. One example was always making sure to join any line that one happened to see, even without knowing why people were standing in line. The reasoning was that, if people were forming a line, there must be something worthwhile to be had. Other strategies included pilfering items from the workplace, parroting the regime every now and again, pretending to go along, making a show of fraternal affection for the Soviet Union without putting any emotion into it—and even going so far as to make scrapbooks attesting to these feigned feelings.

9.1.1 People Have More Than One Opinion

When faced with changes they cannot avoid but which they do not perceive to be to their own personal advantage, people in organizations do not behave in ways that are radically different from those of the inhabitants of the GDR. People go along with changes: always. Although change agents sometimes imply that there is a choice, not going along with the change is not a realistic option for anyone wishing to keep working in an organization that is in a process of change. In many cases, however, people can choose the extent to which they will go along with a change. They are always free to go along with the process just enough to keep up appearances without actually changing.

Robustness allows room for individual freedom, as individuals always belong to groups, communities, or islands of meaning only in part. In the vast majority of cases, people belong to multiple groups, relate to different groups in different ways, and hold more than one opinion simultaneously. This offers employees the leeway to adopt different positions with regard to the change in different places, or to air a different view than might be expected from them, based on their position. Such behavior is not necessarily spiteful or malicious. When my aunt asks me what I think of her new dress, my answer to her is different from what I would say to a friend. By showing different sides of yourself in different situations, you can temporarily slip out of your formal role, voice an alternative opinion, or even "go underground." We observe people using a number of strategies in this way.

9.1.2 Move to the Edges of Reality

It is not possible for everyone to join the communities that have the most influence on deciding the collective reality within an organization. The personal histories of individuals within the organization and their personal views and values can cause them to feel alienated from what the organization has become as a result of the change. Such a shift to another position in the organization does not necessarily involve giving up one's own vision of reality and the behavior that is associated with it. What people tend to do is move toward the edges of the organization and to continue to play the organizational game, only now from a different island of meaning. Their influence in the informal circuit remains undiminished.

> In one organization, a change in management philosophy led to the decision to dissolve a group of middle managers, which had previously played an important role in policy development. This simplified the management of the organization and clarified the chain of command. The group was formally disbanded. It nevertheless continued to exist as an informal group with its own views of reality and its own sphere of influence. Although it is possible to change the structure, doing so does not ensure that people's behavior will change in the process. As long as their behavior has not changed, the robust organizational pattern has not changed either. In this example, the policy group was a very influential group of people, and it had considerable prestige within the organization and beyond. For a group that no longer existed in a formal sense, it retained considerable power, although their influence was now exercised from the sidelines of the organization instead of from the center of the governance structure. The group continued to meet informally and to exert their influence in their personal interactions with others in the organization. The robust organizational pattern remained the same.

9.1.3 Facing Both Ways

Individuals involved in a process of change can decide to support this change in one community, while simultaneously choosing to undermine the change in their own local community.

> The director of a business unit did not agree with a major organizational change, but saw no other option but to fall in line with the vision of the Board. In his dealings within the sphere of the Board, he fell in line with their vision. Among the staff in his own business unit, who shared his views, he took a different line. In this context, he indicated that the change was going to move forward and that everyone would have to go along with it, but

that he would be taking things slowly when implementing the change in the businesses unit. In both of his stories about the change, this director came across as credible, having managed to relate to the views and emotions of the particular group with which he was dealing at the time. By moving between these two stories, he was able to determine his own room for maneuver.

9.1.4   Lip Service

The preceding example shows a conscious strategy for propagating and endorsing different meanings in different contexts. In some cases, the only option open to people is to endorse the new meaning, if they wish to maintain their current positions. In many cases, managers and their advisors would like people to propagate changes and to express their commitment to it openly. On the surface, complying with, surrendering to, or going along with a change despite one's personal disagreement with it bears a striking resemblance to commitment. The major difference lies in the fact that, while people may support a change in their words, their thoughts remain free. Their commitment actually amounts to no more than lip service. As a result, people do not become the motor of the change, but need to be continually prodded into action. They are not motivated from within. After all, if people are not prepared to change their thoughts, their thoughts will keep them from latching on to a change. In the words of a German folk song, "*Die Gedanken sind Frei*" (thoughts are free).

9.1.5   Connect Old Meaning to the New Ideal or Objective

In some cases, new ideals and new goals are effective vehicles to which people can couple their old meaning and accompanying behavior.

> One Dutch municipality was working to create an open and transparent culture of decision-making, both within the administrative organization and among the mayor and councilors. One particular objective of the project was to support the council in this process. The council, however, declared that it already had such a transparent culture, providing the following explanation: nothing ever comes as a surprise in the council meetings, and the decision-making process therefore runs smoothly. When the consultant asked how this was possible, they replied, "Because we arrange everything among ourselves in advance." This was a nice reformulation of what they had already been doing, and it was therefore an effective strategy for avoiding change by claiming that they already conformed to the new ideal.

9.1.6   Have It Your Way, Under a Different Flag

Your personal views on the way things should work are not necessarily shared by the rest of your organization. This does not mean you must sacrifice your personal views. You can continue to cherish them, either in silence or by making a great deal of noise about them. In the latter case, a large part of the organization will probably sigh when they see you coming again with your pet project. There are also opportunities, however, as new goals and activities can help you to win your case. In such situations, changes can help individuals to realize their views more effectively.

> The employee council at one company continued to stress the importance of formulating an organizational position statement and corresponding objectives regarding equal opportunities. These views were not entirely consistent with the priorities of the company director and management. In fact, the eternally recurring subject, the continually repeated plea from the council, and the admonishing fingers that were constantly being waved in the director's face were becoming a source of irritation. This carried on until a golden opportunity arose, when labor shortages prompted HR to develop a new recruitment policy. The employee council struck while the iron was hot, quickly converting their equal-opportunity goals into a proposal for a policy to address the labor shortage by focusing on target groups. Changing circumstances allowed the council to realize their own objective under another heading: the goal of realizing equal-opportunity objectives was replaced by the goal of addressing labor shortages in a strategic way.

9.2　The Miscalculating Employee

Not all calculating employees are equally good at math. In complex wholes, such as organizations, calculating behavior does not always pay off for the individual or the organization. Because of their complexity, organizations are often faced with social dilemmas. That which is in the interest of the individual is not always in the interest of the organization, and that which is in the interest of the organization as a whole is not always in the interests of every individual. When implementing changes, we would like for the changes to be in the interest of the greater whole and, by extension, of the individuals within it. This is obviously not always the case, especially from the employees' point of view. In these situations, we nonetheless assume that employees will contribute to the interest of the organization as a whole. This demands altruistic behavior from employees, but not every employee complies with this demand.

Calculating employees are able to gauge how their own interests stand in relation to the proposed change objective and, based on this assessment, they can choose strategies that assign more weight to their interests. What appears to be the optimal strategy according to an individual's own interests, however, does not always turn out favorably. In such cases, calculated behavior proves to have been a case of miscalculation. Exercises based on game theory can illustrate the way in which miscalculations can arise.

> We once used such an exercise in a meeting involving the mayor and councilors, as well as the management team of a particular municipality. The topic of the conference was cooperation between the members of the two bodies. One element of the program involved a negotiation exercise based on game theory.

In the exercise, the individual participants were free to make their own choices in a series of negotiations. The stated objective was "to achieve the maximum result," although it was not explicitly stated whether the intended result was for the group as a whole or for the individual. If everyone were to choose X, each individual would gain a small profit and the maximum overall profit would be realized. If one of the participants were to choose Y, that individual would make a profit, but would produce a negative overall result. During the exercise, the mayor objected that the objective was not clear, as the interpretation of "the maximum result" was left to each participant. He consistently opted for Y, which meant he was the only one to receive a small profit, while the overall result ended up as negative. In the discussion afterward, he encountered a barrage of criticism from his colleagues for having allowed his own self-interest to prevail. The strategy that the mayor had thought would be profitable for him had ultimately caused him to incur losses in a social sense.

Although it is important for individuals to weigh their own personal interests against the change objective, it is also important for individuals to remember that they are not alone. They must remember that the other individuals involved in this complex field are also making their own choices, whether calculated or otherwise. The effects of one individual's calculations become more difficult to predict as others make their own suboptimal decisions, possibly in reaction to other calculating behavior that they observe around them. This further increases the likelihood of miscalculation.

During a reorganization process, the executive board of a public organization was expanded from two to three members. One of the organization's managers publicly announced his candidacy for the new position. He had a reputation as one of the organization's more effective middle managers, and he had received considerable praise from the executive board in the past for his performance in this position. In an interview with the incumbent board members, he argued that they could not really do without him, that he possessed high qualifications, and that he felt that he had been given the impression he was in line for the position. The two incumbent board members did not consider this coercive attitude consistent with the kind of behavior that they would expect from their prospective colleague. They therefore selected another candidate. The rejected manager was disappointed and, because he had played his stakes so high and so publicly, his star soon fell within the organization. A few months later, he left for a position in a smaller municipality.

9.3 Summary

Calculating employees are aware of their own role in the change process, but they also hold their own views regarding their individual material and emotional interests. Their calculations consist of assessing what change agents are asking of them and what is needed for them to continue to serve their own interests within that context. Employees who engage in calculated behavior appear to go along with changes in order to obtain personal room for maneuver. They are not consistent in the views that they express, keep their own council, use the arguments that are calculated to produce the best advantage, and are continually checking to make sure they do not lose out in a situation. Calculating employees are shrewd opponents, but not always quite as shrewd as they might wish. In some cases, this leads to miscalculations that backfire. Regardless of who suffers the consequences, calculating employees make change a tough process.

Part III

Perspectives

This part looks at two perspectives that are related to unchangeability and robustness. In Chapter 10 we reflect on our own way of looking at things. What makes unchangeability such a difficult subject to grasp? How is it that certain perceived self-evident truths in the fields of management and organizational sciences and change management produce a blind spot for unchangeability? Chapter 11 is, inevitably, about change. What can the preceding chapters teach us regarding change management? And what lessons can change agents learn from the notion that organizations are robust and tenacious?

10

The Blind Spot

In an earlier chapter we described how human shortcomings impose significant limitations on attempts to implement changes. What about our own shortcomings with regard to understanding changeability and unchangeability? As people who study and implement change, we are steeped in a tradition of change, and not one that centers on the absence of change. In order to reflect on what that means, we need the help of a philosopher, which is why we asked René Gude for some reflections on what philosophers through the centuries have said about what can and cannot be changed.

Changes with Lasting Consequences: The Philosophy of Unchangeability

René Gude

Director of the International School for Philosophy in Leusden

Everything flows, nothing remains the same. The reality that we form part of is subject to continuous change. Nature knows no rest: coming into being and passing away is not just the fate of generations of plants, animals, and people, but also of mountain ranges, ice caps, and oceans. On the smaller scale of our daily lives, change is the rule as well. The coffee in front of you will get cold if you do not drink it and will disappear if you do and will thereby be lost, as will the new coat of paint on the garden house, your savings balance, the last quarter's profits, your favorite colleague and our world heritage. In the flow of everything, nothing endures.

People who like change should feel at home in this world, as change seems to be the rule rather than the exception. However, often all that people who profess to love change mean to say is that they could really do with a holiday, not that they like being hit over the head with unexpected events all day long. It is only a very few people who are genuinely favorably disposed toward change, and these have developed the quality of *amor fati*, the almost superhuman acceptance of fate, of which Friedrich Nietzsche considered only a handful of us to be capable.

If we are honest with ourselves, most of us react to change like a warehouse manager faced with the third reorganization in a year's time. A changeable world is unsafe, changeable people are unreliable, and words that change their meaning are unusable. When your partner emerges from a crisis "as another person," this is usually due to another person being involved. Do you react by saying, "Well, I was ready for a change anyway," or do you hurl the dinner service at them?

On closer inspection, every human endeavor, whether it be the formation of your character, your marriage, or your company, ultimately aspires to permanence. Our efforts are ultimately aimed at the sustainable, permanent, intransitive, and, preferably, the everlasting. Deep inside, we want things to stay the way they are and are always looking for survival, continuity, and sustained success. All our plans are forms of mental resistance against the actual change that is continually taking place all around us from day to day. Our personal relationships and material circumstances are subject to such a degree of changeability that it is a mystery where the concept of "unchangeability" comes from in the first place. It is certainly not a reflection of reality. Unchangeability is an idea, an ideal, or an ideology. The greatest single shift in our attempts at attaining happiness in a treacherous world is the invention of the idea of "unchangeability." It serves as a guide in our attempts to halt the changes the world is subject to. Unchangeability is a product of our thinking, a standard that we use to measure the world by. Unchangeability is the triumph of the mind over the facts of our human existence.

I Will Endure

Every organization, whether it be a family unit, company, NGO, political party, school, church, city, or state, is a response on the part of cooperative beings to the reality of continuous change. Every organization is based on an *idea*, which is nothing less than a representation that does not change, even if obstinate reality is continually proving it wrong. For example, every company is based on the unchangeable idea of "making a profit." That idea does not suddenly change after a number of large losses are sustained. It is exactly the combination between the unchangeable idea of "profit" and the deficit on the books, which makes the entrepreneur dissatisfied. The quest for the unchangeability of organizations governs all vision and strategy discussions, all studies on corporate image, mission statements, and branding operations. Whenever organizations desire change, it is the lasting consequences of such changes that they are really interested in.

Everyone is preoccupied with unchangeability, whether it be in terms of household maintenance, the pledge of faithfulness for better or for worse, or a longing for a life after death. Change is usually our natural enemy and only very rarely our friend.

From the Immutability of Substance to Constants in Thinking

For us, there is really only one question that really matters: "What is permanent amid all this change?" Nietzsche called our lifelong protest against our changeable world the "will to power," our unremitting attempt to replace "becoming" with "being." Questions about the substance of things, about their essence, and about the "Being of beings" all come down to the same thing: that which is constant amid all the change. They are no more than strategic reformulations that are meant to bring a satisfactory answer closer within reach.

A quick look at the history of the genesis of philosophy shows that this has always been the big question. The first Greek philosophers from around 600 BC were natural philosophers. The Greek word for changeable nature is "*fysis*" and they called their attempts to gain insight into that changeable nature "*fysica*." The aim of the science they called *fysica* is to discover unchangeable regularities in natural events. The world around us can become predictable if we can get a grip on its harmony, its rhythm, its continuity, its slowness, and its permanence. All science rests on the conviction that there is a knowable unchangeability hidden underneath the chaos of changeable phenomena. Every scientist must have faith in the existence of some kind of order in the world (cosmos) in order to be able to engage with the apparent chaos that surrounds us.

Greek science starts with Thales of Miletus. He was one of the Ionian scientists, traders, and politicians who would not settle for mythological explanations for unexpected events. To them, a solar eclipse was not a singular whim on the part of Zeus, but a standard situation that always occurs when the moon moves in front of the sun, and at moments that can be predicted, moreover. Thales did not just look for explanations for the movements of the heavenly bodies, but also for principles of continuity in the sublunar world. Thales looked for a material element, omnipresent and everlasting, that was present in all changeable phenomena. He thought he had found that element, formulating the statement, "Everything is water." The meaning of this curious pronouncement becomes clearer when one thinks of the fact that water occurs in all the states of aggregation (i.e., solid, fluid, and gaseous), that all living beings consists of water for the greater part, and that the social life of the prosperous city state of Miletus was built on water (Miletus was a flourishing seaport positioned on an estuary). For Thales, water was the unchanging element, which, odorless and colorless itself, was omnipresent in phenomena in changeable nature (the *fysis*) and therefore had to be the starting point for a lasting concept of the larger whole. After Thales, a number of other candidates were considered for the role of the unchangeable element in nature. For example, fire: the substance that can cause the dilution and condensation of water and can therefore be considered to be more "elementary." Earth and air were also proposed, but Anaximander was the one who proposed "to apeiron," that is, "the indefinite" or "the indeterminate" as the underlying element, an immutable substrate from which all the mutable phenomena that we observe arise. For Democritus and Epicurus, this was an important step toward developing a hypothesis of invisible building blocks, called atoms. The natural philosophers took this one step further by giving up on the idea of "matter" and taking the mathematical proportions themselves as the immutable substrate. The world is knowable through its immutable laws, which can be described in reliable mathematical equations that are universally valid. Pythagoras was not interested in a specific, concrete, triangle, but in the fact that *all* right-angled triangles are subject to the same rule, namely that their sides *always* stand in the same immutable proportion to each other. This gave him the idea of the "harmony of the spheres," in which everything hangs together in one large immutable whole. This is what modern day physics, which has long abandoned the concept of "matter," is still in search of. A yet to be formulated *theory of everything* will be purely mathematical and will enable us to predict the immutable laws governing all changes. This means we still will not know what is going to happen, but we will know once and for all how it will happen.

What is most striking is the gradual shift in perspective, from matter to the knowledge of rules. The quest for "immutability of substance" was abandoned and attention shifted to a search for "constants in thinking." Physics has been an abstract science of the mind for centuries. The laws of nature are the product of our thought, which has allowed us to triumph over the facts.

The Spirit Is Strong, the Flesh Is Weak

Heraclitus said that everything we see changes continuously, but the way we see them does not. The logos, our way of looking at things, our frames of reference are unchangeable. Roses fade and ships are lost but the words for "rose" and "ship" endure. Around 500 BC, Heraclitus formulated the fundamental problem regarding the relation between words and things: how is it possible that our concepts have a more or less permanent validity, while the world we perceive is constantly changing? Where did we get our concepts from? Heraclitus is an empiricist is search of lasting insight (logos) into a continuously changing world, constituting a modest triumph over the facts. His contemporary Parmenides took a more radical approach. He was not overly worried by the mutability of substance, but was also particularly delighted by the permanence of our concepts. He had the following line of reasoning: if concepts are more constant than sensory reality, then count your blessings. He is the prototypical rationalist, who starts his quest for a permanent understanding of mutable reality from immutable ideas, such as that of "the triangle" and "the circle." Those ideas are more perfect than any concrete triangle or circle will ever be. The logos is thus perfectly capable of conceptualizing immutable Being. In so far as we perceive changes, this is owing to our perceptions deceiving us, and therefore we should not trust them. This idea constitutes a considerable triumph of the mind over the facts. These two fundamental directions of thought, empiricism and rationalism, are still competing with each other today and the theme that keeps recurring is: what endures, what can we rely on, and what is the one thing that does not change?

Mighty Is the Word

Socrates (469–399 BC) carried on in the line set out by Parmenides. What Socrates does no longer bears any relation to natural philosophy; he is purely interested in ethics and politics. He agrees with the need to look for immutability in ideas instead of matter, but he argues that you then need to literally "make them real" by realizing them. Members of a society can put ideas into practice by converting them into laws, conventions, agreements, moral precepts, mores, and habits. The Greek word for habit is "ethos," and since Socrates the science of habits and the improvement of these has been known as "ethics." In ethics, finding an idea is not the ultimate objective; instead, it is only the beginning of the process. Concepts such as "justice" and "clean drinking water for everyone" are artificial and manmade; they are not products of nature. Nonetheless, these ideas are not fundamentally detached from reality, as they can be realized in a changeable, insecure world. They can be put into practice through the constant, recognizable, reliable behavior of people who agree on the reasonableness of such ideas. If you are looking for stability in the sublunar world, you need to think these values through and attach norms, duties, customs, and virtues to them. By persisting in these habits (ethos) with perseverance, you can realize lasting ideas.

This Socratic idea is also at the core of the philosophy of Plato and it is via him that it has ended up at the heart of Western culture. Immanuel Kant follows the same line of thought: *"An idea is the representation of something perfect which is not yet manifest in reality."* If you can let such an idea motivate your actions, the realization of greater perfection in the world becomes possible. If you take this approach, you are making optimal use of all your possibilities, whether it be in your private life or in an organization. However, the danger—and the charm of it—is that we can realize ideas that we do not derive from nature. It is to that type of creativity that we owe our whole culture, including its attractive and unattractive aspects. In the same way, an organization's staff and management must take the credit for its corporate culture—"even if it's a terrible mess," as the cynic would say. The path to a recognizable character, via an "unchanging idea" and a "practically consistent way of acting," is not without its risks for yourself and your environment. The cynic would say, "The terrorist has an unchanging idea and a consistent way of acting too."

The mind can gain some pretty amazing triumphs over the facts, but this makes it all the more important for us to discuss our ideas. People need a plan, an objective, a duty, an assignment, an ideal—in other words an "idea"—to get them out of bed in the morning. Ideas do not come into being unless we come up with them and they do not accomplish anything unless we make them permanent in practice. Of course one can also retreat into cynicism and be proud of the fact that you have "no idea," but then you would lose out on the chance to give meaning to a teeming universe that you can see the sense of. The permanence of ideas, plans, scenarios, budgets, strategies, contracts, collective agreements, intentions, marital vows, useful taboos, parental responsibilities, manifestos, coalition agreements, conventions, laws, and human rights depends entirely on ourselves. What informs all human endeavors is understanding changeability, uncovering regularities and, taking what is permanent as your guide, bringing something about that would never have existed otherwise.

10   The Blind Spot

René Gude argues that change is inextricably linked to the unchangeable, and that all attempts to change are continually aimed at achieving a form of unchangeability or permanence. According to Gude, a state of permanence is an ideal for which we strive. We are constantly changing the way we go about it in the hope of eventually attaining this immutable ideal. Clearly, just because an ideal is immutable does not mean it cannot be changed. In our rapidly changing society we pay so much attention to change that we run the risk of losing sight of what is unchangeable, and therefore of the possibility that not everything may be changeable.

We notice this attitude in ourselves. While exploring the theme of unchangeability, we quickly find ourselves focusing on changeability. This is partly due to the fact that we have more language at our disposal for describing changeability than for unchangeability, and partly to the fact that we are accustomed to taking this approach. We also discovered that talking about unchangeability with colleagues is not just difficult but can even evoke irritation in some: "Well, if we started believing that . . ."

We went in search of mechanisms that might contribute to this communal "blind spot" for the unchangeable aspects of organizations.

10.1   We See What We Think to See

The information that surrounds us is endless and unbounded, and we can never hope to comprehend it all. Selection is a necessity. We select information on the basis of the information about our surroundings that we already have in our minds. Pregnant women and their partners usually see more pregnant women than they did before the pregnancy. When you have just bought a new car, you tend to see that model much more frequently than you did before. There were obviously just as many pregnant women and examples of that specific model of car around before; it is just that we did not notice them. It is only after a particular category (e.g., "pregnant," "Prius," or "unchangeability") is activated, that the information is actually registered.

This process is reinforced by our tendency to take incomplete information and complete it, mentally, turning parts into logical wholes. For example, consider the common experience of walking along the street behind a person and imagining what they look like based on their hairstyle and gait, only to discover the man or woman concerned is less attractive than we had hoped. In some cases, we even manage to ignore such corrections to our fantasies and hold on to the images we have in our mind. We call this the "halo effect." When we have a positive opinion about someone based on an experience (e.g., "Bill is good with people"), we often extend that judgment to other areas with which we have had no experience (e.g., ". . . so he must be a good dad too"). In even stronger terms:

we see what we expect to see, and we therefore get what we expect to get. The "Thomas theorem" formulates it in this way: *"If men define situations as real, they are real in their consequences."* This also applies to change agents. If people have a great deal of experience in dealing with change, their focus and behavior will naturally be oriented toward change. They observe change, start to act in a change-oriented way (possibly invoking similar behavior in others), and set it in motion, all of which reinforces their existing image of change. We do not have a special sense that we can use to pick up on unchangeability. Moreover, if we do pick up on it, we will usually identify it in terms of a lack of change instead of labeling it as unchangeability.

10.2   Trapped in Our Discourse

The discourse common to the field of organizational science and change management focuses on change and not on unchangeability. This focus is directed by language, which transforms unchangeability into a lack of change instead of a separate characteristic of organizations. Our use of the specific word *"un*changeability" shows how difficult it is to avoid using the discourse and vocabulary of change. The words that we use to refer to unchangeable aspects are nearly always formulated in opposition to change. We have recourse to a range of terms that describe the negative aspects of unchangeability, such as "resistance," "inertia," and "change fatigue." There are also many neutral or positive words that describe unchangeability as a period between changes or after a change, as with "consolidation" and "interbellum." We lack truly adequate words for describing unchangeability.

That imbalance in linguistic richness, which produces a framework in which change appears to have an independent claim to existence and unchangeability is defined primarily in opposition to change, gives rise to a process of reinforcement in which people's attention is influenced by the richness of the available discourse about change. We notice change more quickly and can think and talk about it more easily than we can about unchangeability, for which we cannot fall back on an independent conceptual framework and discourse. Our vocabulary keeps tempting us to think in terms of change and changeability.

10.3   Simplification

The field of change management has an outstanding capacity for producing organizational models. These models are so abundant that there are even books that consist purely of collections of them (Ten Have, 2002). The attractiveness and utility of these models is due to their capacity to stylize our messy reality, making it abstract and intelligible. They help to render complex reasoning comprehensible at a glance. In this book we have also developed a number of models to help shape our own reasoning, and to describe a complex subject in a way that allows us to take the reader along with us in our mental process.

The disadvantage of such models, however, is that they are not merely a reflection of reality, but can actually come to replace reality. They limit our outlook in a different way than our vocabulary does. The aspects that fall beyond the focus of the model are "re-touched" and this can cause us to miss such vital elements as continuity and unchangeability.

> Both of us have been training young consultants for many years now. When working with them, we are confronted with the concepts and models that are used in our common working environment and with their effects. Many of them use the terms "*Ist*" and "*Soll*" in relation to change situations. These German words for "is" and "should be" are used to express the difference between the actual situation and the way it should be. They usually explain this by identifying the *Soll* situation as better and therefore different to the imperfect *Ist* situation. Somehow, this extremely simple pair of terms tempts people into identifying the imperfections of the "Ist" situation in no uncertain terms. If we were to try and make that model even only the slightest bit more complex, by adding a third term (e.g., "endures")—it would probably help people to take a more nuanced view of the future situation.

10.4   The Importance of Change Optimism

Tennis players and those who play other games involving a ball are likely to have had the experience of seeing their own ball as "in," even though the line umpire saw something else. Of course, you can say you saw the ball was in merely in an attempt to convince the umpire, but often it is the case that that is really the way you saw it and that your perception was colored by your self-interest and/or wishful thinking.

What we think we see is strongly influenced by what we "have to" see, as our values, ambitions, and convictions demand this of us. As we indicated previously, this is reflected in the language we use and the models we create. Our interests, ambitions, and desires also play a role in determining our degree of optimism regarding changes. Professional change agents (including clients, change agents, and managers who act as consultants) have a vested interest in seeing changes happen. We are no exceptions to this rule. Change is the reason that consultants and interim managers exist, and many directors and executives are assessed according to the degree to which they have been able to leave their mark on an organization. If the yardstick that your performance is judged by is the degree to which you have managed to realize change, it is imperative that you explain to the outside world all the successful changes that you have achieved. This motivates you to perceive the cases that are successful in more detail than the cases that are not.

The social frameworks of members of organizations also play an important role in this phenomenon. Many managers and consultants today are members of the Baby Boom generation, which came of age with the optimistic attitude toward change that was prevalent in the 1960s and 1970s. They are children of their time, as we all are. Bob Dylan summed up that zeitgeist from 1964 onward in the lyric "The times are a changing."[1] As our convictions color the way we interpret information, the influences that formed them orient a person's outlook on change and strengthen our optimism regarding change.

10.5   More Haste, Less Speed

The likelihood that people will perceive change taking place is also influenced by haste and by our own patience (or lack thereof). In some cases, change agents aim for rapid results, and many clients have similar goals. The tempo demanded by the process of change is one reason for haste. A strong sense of urgency or a need for concreteness on the part of those involved can make quick wins an important objective in the process. This gives people confidence regarding the further course of the process. The second reason is that clients evaluate change agents according to their results. It is in the change agent's own interest to be able to show effects quickly. As a third reason, we can add the psychological need that change agents have for receiving feedback on their efforts: they also want to see results. These are all reasons for framing problems in the most practical way possible, and to aim for solutions in the short term.

In a robust context, however, taking this approach involves the risk of connecting with only a limited part of the robust situation. This can lead to partial and temporary solutions that fail to address the underlying problem. In this way, today's solutions can become tomorrow's problems. It can create a one-sided awareness that focuses on change and not on the larger problem, which has remained unchanged.

10.6   The Logic of Chance

People in organizations undertake countless initiatives each day to improve things, take a different approach, or apply a creative solution to a particular problem. Although there is certainly no lack of change initiatives, we can wonder which of these initiatives will ultimately prove successful and which can be translated into new routines and thus survive. An organization constitutes

---

1 It is interesting to note that, earlier in his career, Dylan had focused on unchangeability, as evident in his song "Forever Young." In contrast with the desire for change expressed by "The times they are a-changing," this song refers to a need for unchangeability instead.

WHY ORGANIZATIONAL CHANGE FAILS

such a complex collection of actions that it is not possible to work out exactly how organic processes of adjustment take place.

In his preface to Chapter 5, Piet Boonekamp explains that chance is at the core of the evolutionary process. Could it therefore be that, as a form of evolution, change is also subject to chance and that, as complex systems, organizations are governed by the laws of chance? Most of us can probably imagine that this is true of spontaneous change initiatives. If it is true for those changes, however, it is probably also a factor in the success of planned approaches to change. Perhaps these owe far less to our efforts than we think.

Leading change and the associated process of selecting change strategies and interventions for change requires reasoning in terms of cause and effect: if we do this, this will happen. We draw links between our actions and what happens next. We must do this, as it is the only way we can check whether our actions are effective. When a change begins, therefore, we assume that it does so as a consequence of the change interventions. Could this also be a matter of chance?

This is an uncomfortable question. People prefer not to see change as the result of chance, as this would diminish the significance of their own efforts. When results cannot be ascribed to our own actions, this produces an unpleasant sense that we are superfluous and that we lack control. John Gray (2002) argues that our freedom of choice is an illusion and that our lives are governed by chance:

> The ancient Greeks were right. The ideal of the chosen life does not square with how we live. We are not authors of our lives; we are not even part-authors of the events that mark us most deeply. Nearly everything that is most important in our lives is unchosen. The time and place we are born, our parents, the first language we speak—these are chance, not choice. It is the casual drift of things that shapes our most fateful relationships. The life of each of us is a chapter of accidents.

Hegel once stated that philosophical reflection serves no other purpose than to eliminate chance. This is also the objective of science. Unable to reconcile himself with the thought that nature was governed by chance, Einstein made his famous statement, "God does not play dice." Unless it intervenes at an opportune moment—we are all acquainted with bad luck—chance is not the most agreeable ally. We suspect that our bias for change results in a blind spot for the role of chance in the intended effect. Moreover, we think that that blind spot leads us to emphasize changeability even more strongly, thereby leaving unchangeability out of the equation.

10.7   Summary

In our desire to make our world ever more suited to our needs, we have developed an extensive vocabulary devoted to changing it, and even a whole academic field dedicated to theorizing and studying change. The discourse that has been developed and the models, constructs, scripts, and paradigms that arise from it also have the effect of weakening our eye for that which does not change. The interest change agents have in realizing changes, the attendant haste, and the need for simplification also help us to preserve this blind spot.

Another aspect of that blind spot is our human need to ascribe meaning to events. This can lead us to interpret effects as the effects of interventions, even though such a conclusion may actually be unfounded.

11

Lessons for Change

After having examined the nature of unchangeability in organizations in great detail, we cannot resist the temptation of drawing a few lessons for the change agent. What can we learn from the foregoing chapters and how can it affect our own approach to implementing change? This is a matter of design, and we are pleased to introduce it with the vision of a real designer: architect and housing association director Frank Bijdendijk. Knowing that he thinks creatively about the function of housing and public space, as well as what they mean to people, we asked him to write about the changeability and unchangeability of buildings.

The Yin and Yang of Our Built Environment

Frank Bijdendijk

General manager of housing association Stadsgenoot in Amsterdam

I have been the general manager of an Amsterdam housing association for twenty-five years now. As is the case for all housing associations, the raison d'être of this organization is the need to provide people with a place to live. That is why I am particularly interested in what the inhabitants themselves think of the housing they are provided with, and what are the things that really matter to them. Looking at developments in the Netherlands since the Second World War, I cannot but conclude that some big mistakes have been made. The most important mistake lay in the fact that governments and professionals for many years thought they knew what was best for citizens, and what it was they needed. This resulted in the monotonous, monofunctional, and impoverished post-war residential neighborhoods we see today, many of which are now known as problem neighborhoods. Also consider the style of urban planning that separates functions on a macro scale. What we have created is cities of isolation. In October 2005, riots broke out in a number of French suburbs. Why there? You only need to look at the photographs to see why. The alienating apartment blocks built on a huge scale, their facades signaling the poverty of the people that live inside: the palpable sense of neglect. You can bet that the people responsible for designing these estates do not live here themselves. These images touched a nerve in me and have got me pondering what possibilities there are for a built environment that does meet the needs of people of flesh and blood, and thinking about a certain sustainable quality of life and housing. This has given me the courage to write about the topic here. It has given me the courage to say something about the characteristics of such a built environment and, in the framework of this book, about the analogies that can be drawn to organizations and companies, which are also environments in which people work and produce. For the past twenty-five years, I have wanted to bring these two worlds together: the world of capital and the world of day-to-day human needs. In my work I have invested millions, but I have also been received in people's homes.

There are at least two similarities between our built environment and our organizations: both are constructed by people, for people, and both require capital-intensive investment. That is why sustainable quality is of such importance in both cases, for users as well as for the necessary return on investments. Can these two goals be reconciled with each other? In order to answer this question, let us first take a closer look at our built environment, thereby asking ourselves how sustainable quality is created.

My suggestion is to take the concept of sustainability as our starting point, in its sense of the durability of systems. Sustainability relates to our environment, the world in which we live. Our built environment definitely forms a part of this. The question we are faced with is how to render it sustainable. My answer is based on the view that the durability of systems is fundamentally connected to the degree of responsibility that those systems evoke in people. Time and time again, history has shown us how damage to our living environment has been the result of irresponsible human behavior. Such behavior can stem from ignorance, from nonchalance, or from the fact that people simply do not feel any connection to the relevant component of the living environment. This last motive provides us with a starting point for our consideration of the built environment. When it comes down to it, people want to feel they can relate to their environment, whether it be a landscape, a forest, a beach, or the place where they live or work. This need for connection is something of all ages and all places, and constitutes a human desire to relate to something or someone that is larger than oneself and exists on a larger timescale. The best proof of this is probably the meaning religion has had for billions of people throughout the ages: The Latin verb "*religare*" quite simply means "to connect." Another proof can be found in the way in which peoples that are close to nature deal with their living environment. What are the qualities that allow people to relate to buildings, public spaces, villages and towns? Let us start with buildings, first of all.

If we compare them to people, buildings have a couple of conspicuous characteristics: they are larger than people and generally last longer than a human lifetime. Secondly, they have an inside and an outside: You can enter buildings but you can also look at them from outside. Inside, buildings fulfill a use function; on the outside, they are part of public space. On the outside, they can be attractive, luxurious, and beautiful or, in the other extreme, shabby, alienating, and ugly.

The difference between inside and outside plays an important role in the thought processes of professionals in the building industry, including architects and urban developers. Their basic assumption is that the outside should reflect a building's function on the inside. "Form follows function" is the classic maxim taught in architecture courses at universities. However, in reality, actual buildings take little notice of this architectural rule in the uses to which they are put. A building's outer form and inner function need not be related to each other at all: just consider how often the functions of Amsterdam's canal houses have changed throughout their 350-year history, while retaining their appearance of merchants' residences. It is this difference between the inside and outside that we should look to if we want to uncover the way in which people relate to buildings, and the kind of relationships that produce sustainable quality.

Firstly, we will look at sustainable quality of the inside. This requires a building's use functions to connect to people's needs and emotions. How can we achieve this? After all, no two people are the same. Moreover, a person's needs change continually during their lifetime, along with their changing circumstances. They also change because the society they live in is continually changing—and changing at an increasingly rapid pace. This is driven by technological developments, as well as changes in types of labor and housing needs. Finally, we live in a society made up of emancipated and well-informed citizens with a fair degree of financial independence. We really have no choice but to take the fact of changeability, combined with people's emancipation as our starting point. The implication is that the interiors of buildings need to be able to accommodate continually changing functions and that the users must be able to decide what these functions should be. This constitutes the ultimate form of sustainable quality with regard to the insides of buildings. Essentially, it combines a capacity for accommodation with giving users the freedom to adapt their living and working space to their needs.

There is little to argue with there, in my view. Who is not attracted by the possibility of changing the layout and contents of one's living- or workspace in no time and as often as one likes, adapting it to new needs and requirements? It seems to me this would be a particularly effective way of enabling people to feel that a space is truly theirs.

Let us turn to the outside for a moment. Of course it protects the inside against the elements and all manner of unwelcome influences from outside. However, as we have already seen, that outside also helps to shape public space. Moreover, through its design, the choice of materials, and the detailing, the outside has a considerable influence on the attractiveness of both the public space and the building. This is a function that the outside fulfills not just for a building's individual users but above all for the public, the collective formed by the people who make use of the public space—that is, for the residents of a village or town, the inhabitants of an area, neighborhood, or street. We all know from our own experience how certain outsides can have either a positive or a negative effect on our feelings. I already mentioned the example of the French suburbs. In the Netherlands, a clear example of outsides exerting a negative influence is the case of the Amsterdam suburb Bijlmermeer. This development existed for only thirty years: Having been delivered around 1968, demolition work started in 1998. How could this happen? Not because the buildings were derelict, because the apartments were too small or because there was no space for parking, driving or recreation—or even, as some people suggested, because the "wrong sort of people" were housed there. The reason was quite simply that the Bijlmer project had an alienating and intimidating effect on people, that it proved incapable of appealing to people's affections, and that people were not able to relate to the environment in a personal way. There are plenty of examples of environments that have the opposite effect. Just take the center of Amsterdam: who does not enjoy strolling along its canals? What else can explain the growing number of people fighting to protect large parts of our cities and villages? Why else was Warsaw rebuilt as a near carbon copy after nearly 85% of the city had been destroyed in the Second World War? The reason behind all of this is that people feel a connection with these environments, identify with them, and therefore feel a sense of responsibility. Our built environment, in its function as public space, can be something that is dear to us: something precious. And there we have hit on a second sustainable characteristic of buildings and the built environment: the personal attachment they inspire in us, that is, their preciousness. Again, this is something I think most people would agree with. After all, do we not share the view that beautiful streets, squares, parks, neighborhoods, villages and towns should be preserved, simply in order that we might enjoy them?

The ability to inspire affection and a flexible accommodation capacity: these are the two things that really matter. Emotional value must be linked to unchanging forms and materializations, and flexible accommodation capacity must be linked to continually changing functions within those unchanging forms. Both these factors relate directly to our lasting human needs and emotions. That is why they add up to a sustainable quality of our buildings and our built environment. You cannot separate the two: the two qualities of changeability and unchangeability are as inextricably linked as yin and yang.

It is on this dual principle that the Solid is based, an innovative building concept used by Het Oosten. What is characteristic for Solids is that the shell and the inbuilt components of the building are disconnected from each other in all respects: with regard to technology, life span, ownership, and personal choice. Het Oosten leases the shell and the user decides what to do with the inside. The affective value of the outside is strengthened through the use of materials that age in an attractive way, through abundant detailing, and through a majestic entrance which connects it to its environment. The shell's capacity for flexible accommodation is great owing to the generous proportions of its stories, large column-free bays, floors with increased load-bearing capacity, with its connections to services (water and electricity) at many different points and its optimalized acoustics and energy consumption. These Solids require very high start-up investments. However, these investments are amply recouped thanks to their sustainable qualities, which are manifested in low maintenance costs, low conversion costs and the limited leasing risk. Moreover, the user takes all the decisions with regard to way the inside is fitted out. This is possible due to the way the inside and outside function independently and, significantly, to the great degree of freedom provided for in the zoning plan, which means the possibilities are endless!

We began by pointing out there are at least two similarities between our environment and our organizations. Is it possible to take the line of reasoning we followed with regard to buildings and extend it to organizations? Is it not true for organizations too that changeability exists (on a smaller scale) within unchangeability (on a larger scale) and that these two qualities are as inseparable as yin and yang? After all, organizations are also made by people, for people. Again, the answer seems obvious to me, and we can thereby identify this as a third important similarity.

If we take all of this to be true, a fourth similarity remains between our built environment and our organizations. This relates to the capital-intensive investments we mentioned at the beginning. It goes without saying that these must be recouped. With regard to buildings, this generally happens in two different ways. Firstly, through the annual net cash flow produced by a building for its investors, which is basically the rent collected minus the operating costs. It is the relation between that cash flow and the investments made that determines the so-called direct return. However, the buildings' scale and their long lifespan means that they also represent intrinsic value. In fact, the value of buildings with a sustainable quality even increases from year to year, thereby producing capital growth. It represents an addition instead of a deduction. That increase in value, divided by the funds invested, is called the indirect return. Of course, what investors are interested in is the sum of the direct and indirect return. This needs to be as high as possible. We can ask ourselves what would produce the higher return: buildings that people love and feel a connection to or the opposite? The answer is that the emotional value people place on buildings has a direct effect on their economic value and results in a higher annual cash flow.

Therefore, "appreciation" is a term that has two sides to it: emotion and economics. Once again, these are two aspects that are inseparably connected. It follows that sustainable quality of buildings, translated to their accommodation capacity (inside) and their affective value (outside), also result in a higher total return. A beautiful, attractive building—a Solid, an urban palace—which is permanently usable, and which can accommodate any function; such a building can always be leased and will therefore produce a steady cash flow. Moreover, it will also increase in value, owing to the fact that people relate to them. This brings us to the fourth similarity between our built environment and our organizations: organizations that display these same kinds of sustainable qualities can also be sure to produce an annual profit as well as increase in value. In this light, sustainable quality is not just a good thing for a building's users or an organization's employees, but also for the people who invest in them: a case of yin and yang to the second power.

11 Lessons for Change

Unchangeability lies at the heart of this book. We must nonetheless wonder about the potential significance of the insights we have gained for the change agent. In this chapter, we present a number of considerations or lessons that can be derived from the preceding chapters.

11.1 The Value of Unchangeability

Frank Bijdendijk attributes a quality of preciousness to buildings. To Bijdendijk, preciousness is an important characteristic, as it makes people feel at home. The stable characteristics of robustness are also precious to people. They are the characteristics to which people can relate, and which allow individual acts of sensemaking to become intertwined with the collective sensemaking of the organization. Individual meaning is intertwined with collective meaning and the two converge, at least in part. The world of the organization is also partly the world of the individual. It is this connection that creates feelings of attachment on the part of individuals: attachment to routines and behavioral patterns, as well as to their organization's ideals, values, and convictions to which people feel they can relate.

Another feature of that stable, trusted robustness is that it offers individuals freedom. It is because people know their specific contexts that they also know what room these contexts offer for maneuvering. This is analogous to moving to a new living environment after having lived in another place for a very long time. In your old surroundings, you knew your way around, you were familiar with local habits, you knew the people and how they reacted to you, you knew who you were in that community, you knew where to buy the best strawberries and where the best places were to get lunch. Moving around in that environment and being free to take up your own space felt very natural and familiar to you. Having moved to a different place, however, all of the knowledge you had accumulated is lost. You do not know where the nearest drugstore is, you are not familiar with your new neighborhood and the local mores, and you have not yet discovered how to move about within and make your way through that new community. It takes time to learn these things. This also means losing some of your freedom: you must retrace your boundaries and this will sometimes prevent you from doing things the way you are accustomed to doing them, as you need to start over again and learn how things work in your new context.

In an organization that had just undergone a large-scale merger between three parties, an employee once told us how, after having worked with one of the companies involved in the merger for many years, he was so familiar with the customary routine, with the way things worked around there and what was expected of people, that he felt as comfortable as a fish in water, moving freely within a familiar context. In the new organization he did not know his way around and he was faced with much that was unfamiliar. This interfered with his freedom of action and it made it difficult for him to claim space for himself. Because he did not know his way around, he felt he was far less effective at carrying out his professional duties, and that he was unable (particularly at first) to exert an influence on the organization, his own work, and his cooperation with colleagues. It was only when he was able to bring the organization's routines and his own actions in line with each other to such an extent that they became habitual, that he felt he was effective again.

Managers, directors and change agents are not always aware of how precious robustness can be to people. They may use fine words in long-term strategies, which proclaim that an organization's employees form its true capital, and attempt to connect real consequences to the results of the latest employee-satisfaction survey. Despite all this, that which the average employee considers precious is usually subordinated to that which is considered strategically important. At the same time, we suggest that the change will create a new preciousness to treasure if they go along with a change. We give them the impression that the change will offer advantages to them as well, and that these advantages will compensate for the precious qualities that are lost in the process.

Tenacity "protects" organizations against all too aggressive forms of change mania. It protects robustness against changes that are simply too ambitious. When an organization reacts to a change with tenacity, this allows its old meaning to be maintained and its core to remain intact. This allows those aspects of an organization that people value to be retained. For the change agent, tenacity functions as a kind of signpost. It indicates that the energy within an organization is primarily focused on consolidating robustness and not on bringing about the proposed change. Tenacity can be compared to resistance, in terms of its value for change agents. Resistance is usually taken as an important sign indicating that people are not capable of working with the change. Tenacity functions in a similar way, only at the level of the organization as a whole.

While attempts to eliminate resistance often focus on taking away concerns on the part of the people involved, thereby allowing them to go along with the change, we think that tenacity requires an analysis of the change, and any aspects of it that threaten to destroy something that is of substantial importance to the organization.

11.2 The Governance Paradox of Robust Systems

Robust systems construct themselves from within, including their own governance mechanisms. These mechanisms are complex and layered—there is no single person at the helm. A considerable part of governance rests in the values, convictions, and behavioral patterns of people within a complex series of interconnections. The system largely regulates itself. At the same time, it has a tendency to close itself off, focusing too much on the internal world and too little on the external world. The first characteristic of robustness (i.e., self-guidance) requires a form of governance that facilitates and stimulates the internal self-regulating mechanism. The second characteristic (i.e., a tendency to become closed off from the outside) demands firm governance interventions (including those from outside of the system). This requires those at the top to continue to govern where necessary, even if they are not the only ones to govern.

Excessive dependence on the power of the robust system's internal self-governance mechanism can blind people to the necessity of "external" intervention. Conversely, excessive dependence on "external" governance does not do justice to the self-governing nature of the system. The person who governs an organization without having a monopoly on governance must be capable of allowing robustness to shape and maintain itself, while simultaneously being prepared to intervene whenever an imbalance emerges between stability and dynamism.

11.3   The Human Condition

With their average qualities and capacities, average people find their place within robust systems that subsequently provide them with a structure to support their actions. Imperfection and shortcomings are inherent to the system of organizing. In this way, human capabilities and robust contexts reach an internal balance. The degree to which a system allows room for change within a system depends upon the capacities of the people involved. In the field of change management, that factor tends not to be taken into account. The most common image that is used to describe how people change features employees who want to change as long as they are involved in the change, who can change if they are allowed to learn the required skills, who are capable of great achievements, and who are focused on self-development. We do meet people who conform to this image in our daily work. Nonetheless, we also see many people who do not conform to it, whether in whole or in part. The diversity of people with regard to their capacity for change is only rarely considered when change strategies are being designed. We often look at the personal qualities of individuals and what they can and cannot do, but we never (or only very rarely) consider the degree to which they are able to go along with a change. Moreover, we do not ask ourselves often enough whether the population as a whole has the potential to change according to the ambitions of management. As a result, the capacity of people to contribute to a change is often either over- or underutilized.

11.4   The Sense and Nonsense of Change

If an organization's robustness does indeed have that value that we attribute to it, it raises the question of why one would want to change an organization in the first place. After all, a robust organization combines the capacity for staying true to itself with the capacity for changing adequately in response to its changing environment. These capacities alone, however, cannot guarantee the continued health of an organization. As we have already seen in Chapter 5, "Pathological Forms of Robustness," robustness does not always maintain its own balance, as it can tip over into inertia or an excessive dynamism. The lesson we can learn from the authors of this book's introductory essays is that change is activated either because of incisive external changes or because, in the course of the organizing process, those involved happen upon an adaptation that works better. Consider Louise Vet's description of the parasitic wasps, which go to a different plant only if the preferred plant of their host is not available. Interventions in linguistic usage fail unless they are anchored in something that has already developed in practice.

As a very general rule, organizations that are excessively dynamic require interventions that are aimed at restoring stability, whereas organizations that have lapsed into inertia require interventions that generate dynamics. Following the line from the introductory essays and applying it to the change repertoire, we must understand that, although it is important to keep sensemaking processes healthy, it is equally important to break out of patterns that have become dysfunctional. Moreover, we should always consider the abovementioned shortcomings and the governance paradox of robust systems.

11.5   The Perspective of Change Management

Change strategies that provide room for human limitations are strategies that take the robustness of organizations into account. Although robustness is anchored in the meaning and sensemaking processes that have been created within organizations, opting for change strategies aimed at facilitating sensemaking processes and stimulating the organization's own capacity for change may not always be the right choice.

11.5.1   *Flexibility Within Robustness*
Robustness is a complex phenomena made up of sensemaking processes, the actions of people, behavioral patterns, and constructs. Understanding what needs to be changed and which strategy is best suited to achieving that change requires a thorough analysis of all of these facets and their relationship to each other. Such an analysis should produce the following insights into the diagnostic phase of an intervention:

- insight into the degree to which sensemaking processes are aimed at maintaining stability or at creating adaptations and changes
- insight into the degree in which behavioral patterns and people's actions are intertwined and into the amount of room that is allowed for diversity with regard to action
- insight into the way in which interpretations of reality are directed by constructs and into the amount of room that is allowed for ambiguity.

Insight into these aspects helps to create an image of an organization's capacity for change. Van Oss (2002) has the following to say on the matter:

> In order to be able to adapt to changes in their surroundings, it is important for organizations to pick up on signals from those surroundings and act upon them. That lies at the very core of organizing. This makes the organization sensitive to changes in its environment, which allows it to adapt and adjust more easily to the often changing demands of the environment. When the process of organizing proceeds in a healthy way, there is room within the organization for new meaning, (or for the co-existence of different meanings), retentions are not enforced, and people become actively involved in the change rather than reacting defensively. When the process of organizing proceeds in a closed manner, however, its reality is determined by the existing retentions (including cultural retentions) held by individuals and groups. The organization fails to pick up on signals coming from its environment. For all intents and purposes, therefore, the organization is closed off to information from the outside world.
>
> Weick lists a number of aspects that are important for keeping the process of organizing alive:
>
> For the cognitive aspect of sensemaking it is important for people to be:
> * aware of their sensemaking processes and of their limitations;
> * capable of coping with ambiguity and uncertainty;
> * able to cope with different realities.
>
> For the action-related aspect it is important for people to be:
> * capable of searching for realities that are broader than their own through dialogue with each other;
> * able to incorporate variety and diversity;
> * capable of reacting to changes in a flexible way.
>
> For the social aspect it is important that:
> * the organization provides space for direct contact and interaction, (including with the outside world) for as many employees as possible.
> * there is a sense of mutual respect and trust.
> * there is room for experimentation and making mistakes.

From a social-constructivist perspective, a change can either focus primarily on sensemaking processes, or primarily on constructs (or the changing of constructs). Which focus is most suitable depends on the degree to which an organization's robustness is closed in on itself, the amount of room that organizations and people have left for alternative constructs, and the extent to which the organization displays healthy sensemaking processes.

11.5.2 The Function of Tenacity

As we described in Chapter 6, changes drive a wedge between people's behavior and the internal robustness of an organization. An organization in this position is likely to react by standing its ground in an attempt to reconnect behavior and robustness. Tenacity can be a functional reaction on the part of an organization when a change threatens to damage its robustness. However, tenacity can also be dysfunctional, when it stands in the way of a change that is necessary in order to keep an organization's robustness healthy and balanced.

Interventions are based on, utilize, and facilitate the organization's own capacity for change are likely to evoke only a limited degree of tenacity. Tenacity will be amplified if the change affects the core of the organization's robustness. In a situation that is excessively closed, tenacity can constitute a dysfunctional reaction. If analysis shows that an organization's robustness is too rigid, and that sensemaking processes are excessively focused on maintaining stability, effecting any real change will require "external," enforced intervention. This is likely to increase the tenacity of the reaction to change. In order to effect real change, however, it is necessary to evoke such a reaction.

11.5.3 Intervention Framework

Summarizing the insights described above into a quadrant produces the following framework for intervention:

Table 11.1 Intervention Framework

Focus Method	Changeability of sensemaking processes	Robust side of sensemaking processes
Facilitating	Creating preconditions for sensemaking 1	Offering new meaning: sensegiving 3
Coercive	Organizing reflection 2	Enforcing new meaning 4

The four quarters of the quadrant provide direction for involving sensemaking processes and built constructs in a process of change:

1. Creating preconditions: if there is a healthy balance between change and stability, and if an organization's robustness is dynamic instead of closed, a change process can be aimed at creating preconditions for organizing new meaning. Merely inviting groups and teams to consider a certain issue may be sufficient.
2. Organizing reflection: in situations in which the process of renewing meaning no longer takes place of its own accord and in which the creation of preconditions is not sufficient, change strategies can be aimed at consciously organizing reflection on existing mental models and patterns of action in order to stimulate the process of creating new meaning. Organizing peer

consultation sessions, internships, job rotation, or a visit to a museum can help people to redefine their own meaning.

3. Sense-giving: if meaning has become entrenched to the extent that the stimulation of sensemaking processes no longer works sufficiently, one option is to extend meaning through sense-giving. Gioia and Chittipedi (1991) describe an iterative process of sense-giving and sensemaking in organizations. They argue that people arrive at new meaning only when they are supported in their sensemaking by someone who presents them with alternatives by way of sense-giving. This refers to the activities of change agents and managers that are aimed at offering attractive new meanings.

4. Enforcing new meaning: when an organization's robustness is so closed off that behavioral patterns and constructs are no longer adaptable, it is necessary to enforce new meaning. A decision can create a new reality as a given fact, thereby making the existing meaning untenable.

In the book *Zinnig ondernemen* (Meaningful business), Van Diest (1997) provides an outline of developments within the field of organizational theory, showing how the discipline developed from its foundation in strongly positivist views of organizations to theories informed by a more social-constructivist framework. Van Diest rightly argues that, although the latter theories are increasing in importance, the theories that were developed before continue to have an important function and role in organizations. The quadrant also demonstrates the importance of both organizational perspectives to change management. While the first three quarters indicate ways of influencing the sensemaking processes of people in organizations, the approach indicated in the fourth is more consistent with classic planned-change repertoires.

11.6   There Is Not Just One Change

The intervention framework presents four change configurations. They can nonetheless be implemented in many different ways. And, we feel it is important that they should be implemented in many different ways. Attempting to divide the world of organizations and organizational change into four categories would involve too great a reduction of the complexity of organizations. The governance paradox and the reality of human shortcomings that we have described in this chapter play an important role in this complexity, even if changes are designed according to the actual situation of an organization's robustness.

Changes are often conceived at the highest levels of an organization and "rolled out" from above. A robust system consists of robust subsystems with their own realities, behavioral patterns, and views regarding the change that is being implemented. If an organization is a constellation of islands of meaning, it is possible that the effect of an intervention conceived at one level will be different as early as the next phase or the next level down. An intervention can also have varying effects on different components within the same hierarchical level. Visualizing the organizations as a linking-pin structure, each link represents a different change, based on the robust characteristics, the capacities of people, and the external force that makes the change necessary. The term "roll out" does not do justice to the complexity of organizations and the combination of changes that may be required. It invokes an image of a carpet-layer installing wall-to-wall carpeting: a thick roll of identical material. This does not reflect the customized working methods that are required in order to realize realistic change.

In our view, rolling out a change within a robust system requires a succession of other changes. This means that a separate assessment should take place for each component, examining what is necessary given the situational balance between robustness, the actions of people, and the relevant change strategy. For example, changes in the production department may involve a different type of robustness than do changes in the sales team, and they may therefore require a different approach. Change agents must thus be able to connect these factors and translate them into changes that are customized and suited to specific locations within organizations.

Each component in Figure 11.1 has its own character and therefore requires its own specific change strategy.

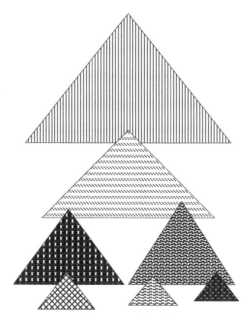

Figure 11.1   Differences within change.

11.7 A Sense of Possibility and a Sense of Reality

Limitations to implementing change are not only present in organizations; they can be found in change agents, as well. In *The Man without Qualities*, written in the early 1930s, Robert Musil (1995) describes people who possess a sense of possibility and a sense of reality:

> To pass freely through open doors, it is necessary to respect the fact that they have solid frames. This principle, by which the old professor had always lived, is simply a requisite of the sense of reality. But if there is a sense of reality, and no one will doubt that it has its justification for existing, then there must also be something we can call a sense of possibility.
>
> Whoever has it does not say, for instance: Here this or that has happened, will happen, must happen; but he invents: here this or that might, could, or ought to happen. If he is told that something is the way it is, he will think: Well, it could probably just as well be otherwise. So the sense of possibility could be defined outright as the ability to conceive of everything there might be just as well, and to attach no more importance to what is than to what is not.

Many professional change agents can probably identify with the image of a person with a well-developed sense of possibility. It is impossible to change anything without being able to think in terms of alternatives to that which already exists. At the same time, what Musil's description illustrates is how a well-developed sense of possibility can lead to the excessive relativization of the world. Before we know it, we continue to come up against the sturdy frame of existing reality. It is important for change agents to continue to distinguish between the reality of the robust organization and the virtual reality that they are proposing to achieve through the change. The trick is to respect this distinction in such a way that it enables a connection to arise between the existing and proposed realities.

Recognizing this does not imply that change is impossible. It does, however, have implications for the level of ambition and the focus involved in changes. Not everything is changeable, and changes require interventions that are more precise and conscientious, in order to enable people to preserve what is precious to them, and to change that which really needs to be changed.

Conclusion

The Writing Process

This book has been written by two people who carry out their daily work in a world of change, a world in which changeability is never really questioned. There, we see how difficult it sometimes is for organizations and people to change, and how at times change even seems impossible to achieve. Over the years, and against the tide of change optimism, we came to develop a kind of change realism. When we reflected on our experiences, we could clearly see how often the results of well-intentioned changes failed to materialize or how changes were only partially effected. It seemed as if organizations or parts of organizations were unchangeable, and that organizations had certain mechanisms for keeping change at a distance.

This awakened our curiosity and started us on our quest to look for the unchangeable side of organizations. At the beginning of 2006, we organized a small conference that brought together managers and organizational consultants. We asked a number of scientists and scholars from other disciplines to enlighten us about the role unchangeability plays in their own fields. Taking these introductory lectures as our starting point, we talked about what relevance and value these stories and insights might have for the field of organizational and change management. As the conference came to an end, we personally were left with two things that had struck us in particular. Firstly, the event had really helped us to gain more of an insight into the blind spots within our own professional field. The second point that struck us was how difficult it was for managers and consultants to talk about unchangeability. It was not an easy topic to explore with people who continually work on change in their professional practice and have been trained to come up with solutions for every problem relating to change—and we are by no means excluding ourselves from this group.

When we told people that we were investigating the unchangeable side of organizations, the reactions we received can be divided into three categories. The first type of reaction was one of recognition and relief. It gave people the language needed to verbalize something they observed but could not easily define on put a name to. During the aforementioned conference, for example, one of the managers who attended saw that the power that managers have to change things is in reality very limited. It reassured him to understand that his ambition for change had up until then been excessive in relation to the organization's changeability. Another category was formed by reactions characterized by denial or even anger. For the people who reacted in this way, change was the be all and end all. If things failed to change, they either felt this must be due to a wrong approach, or they saw it in terms of a temporary blip which could be conquered. A true believer does not give up easily. The third category consisted of reactions on the part of colleagues who questioned the commercial sense involved in investigating the phenomenon of unchangeability; after all, we make a living from doing things that *are* possible, right?

An Unforeseen and Paradoxical Intervention

A number of people have remarked upon the irony involved in simultaneously writing a book about unchangeability and working on changeability in practice. When we took our leave from one organization where we had carried out an interim assignment, our client told us how, after news had spread through the organization that we were working on this book, everyone had done their level best to prove the authors wrong by showing them that their organization did indeed have the capacity to change.

This brings us very close to what our intentions have been in writing this book. Our aim is not to demonstrate that nothing can change or that we are incapable of changing anything in organizations; it is our aim, however, to show that there are parts of organizations that are largely or wholly unchangeable. Being aware of and respecting such manifestations of unchangeability can contribute to a more pleasant, more fluid, and more effective practice of implementing change, because the energy put into the process can be expended on things that can actually be changed.

From Resistance to Recognition

We started our journey in search of unchangeability from feelings of irritation and of having failed in our own professional practice of implementing change. Why are many of the changes we envisage not more successful? Why is it that, a few years on, we often observe that less has actually changed than we thought was the case when we left? Why do many changes proceed so sluggishly? Our irritation was based on the fact that all of this used up a great deal of energy, and our feelings of inadequacy resulted from the high expectations many of the clients seem to have, combined with the fact that some of our colleagues did have success stories to tell. We immersed ourselves in theories about resistance and the different ways of dealing with it, and we saw how resistance often also took the form of our own resistance to resistance. After all, the resistance that change agents attribute to others often tends to call up feelings of irritation and resistance in themselves in turn.

Gradually our interest, which stemmed from negative experiences, turned into a fascination with the phenomenon of unchangeability as such. We became intrigued by the fact that so many things turn out to be unchangeable. We started to wonder why, when so many changes fail, so little research is done into the unchangeable side of organizations. Another thing that struck us was how, in assessments of failed changes, people tend to automatically focus on the approach taken ("If you had taken my advice, things would have turned out differently . . ."), while failing to take into account the actual changeability of the "object" that is to be changed.

During the writing process we were repeatedly confronted with the degree to which we ourselves were afflicted with this "focus on change." While writing it was often difficult for us to keep ourselves from straying from the side of unchangeability. This also explains the struggle we experienced in trying to define what changeability, once intuited, actually is: what is it, where does it reside, and what does it consist of? We regularly caught ourselves reverting to the familiar ground of discussing change instead of unchangeability. It was a bit like trying to keep a roly-poly toy on its side. Trying to consistently think in terms of unchangeability proved to be real brainteaser. The chapter titled "The Blind Spot" was written in an attempt to uncover the background to our own restricted outlook.

Eventually it dawned on us that, while unchangeability may often pose a problem to change agents, for the organization involved it often holds the answer. As we proceeded, we developed an ever-clearer picture of organizations that are imperfect like everything else, but which have earned their spurs and reason for existence. It also became clearer how change agents confronted these organizations with their illusionary aims for perfection.

Do we, as professionals change agents, really understand what this is actually about? We observe ongoing public debates about failed changes in the education system, an excessive insistence on innovation in the health care sector, and the managers who seem to only get in the way of people trying to do their work. We observe how public sector organizations go through one reorganization process after another and how each newfound form of dissatisfaction appears to be translated into yet another proposed process of change. We see how the authoritative Henry Mintzberg (2004) repeatedly criticizes MBA programs for producing graduates who see the world in terms of business models and have no real connection to the daily reality of work. We take note of Chris van der Heijden's (2007) pamphlet criticizing the "culture of managers" in the Netherlands, which he likens to the regime of the Soviet Union, asking himself how it is possible to free ourselves from its dictatorship.

We feel increasing sympathy and understanding for the position organizations and their inhabitants are in, as they arm themselves against attack, neglect, destruction, ignorance, and the manic pursuit of change for change's sake. We are coming to see more and more clearly what a blessing it is that organizations do not allow everything about them to be changed. That capacity for unchangeability is a strength. The robustness it produces is rooted in the past and is therefore the key to the future. Moreover, the reaction we are confronted with as change agents, which we have described as tenacity in this book, is actually the manifestation of an organization's vitality, strength, and value. We should really be thankful that so many change initiatives fail!

Our book is now finished. It is caught between covers and therefore unchangeable. However, we hope the process of thinking about unchangeability keeps moving and developing for a long while yet.

Leike van Oss & Jaap van 't Hek
Aldeboarn, Maarssen, Rouzède, Ruinen, Schoorl, Utrecht, 2007–2008

Appendix A: Social-Constructivism

Social-constructivism is a philosophy that is based on the fundamental assumption that reality is not something that is given but is constructed by people collectively, through interaction. Social-constructivism is also known as social-constructionism. The latter term is more commonly used in social sciences and philosophy, and the former more commonly in psychology. Social-constructivism attributes a central role to the process of sensemaking. These sensemaking processes allow people to construct meaning and systems of meaning. I this book we mainly refer to theories developed by Karl Weick, Dian Marie Hosking, Ian E. Morley, Thijs Homan, and Wim de Moor. These theories also feature prominently in this appendix.

Reality

De Moor (1995) makes a distinction between constructionists[1] who do assume there is an external reality and more radical constructionists for whom there is no such thing as a material, external reality, but who see reality as something that is entirely constructed through processes of social interaction. In both cases, what is central to the sensemaking process is the way in which people more or less actively interpret the world around them and give it meaning.

Individual and Collective

De Moor also makes a distinction between individual constructionism and social-constructivism. Individual constructionism relates to the way people construct individual perceptions, while social-constructivism describes the way in which people, as social creatures, have access to "common sensemaking" in interaction with each other and together develop social realities (common senses).

---

1   De Moor uses the term "constructionism," where we have chosen to use the term "constructivism."

Individual and collective sensemaking processes are intertwined. Both individual and social sensemaking comes into being through human interaction. That interaction results in individual images of realities and a collective image, in which a converged image of reality has been created through interaction.

Individual Sensemaking

According to De Moor (1995), individual sensemaking is a process that shifts between thinking, feeling, and acting. In order to describe individual sensemaking, De Moor uses a so-called ABC model, which is based on the rational emotive behavior therapy developed by Albert Ellis.

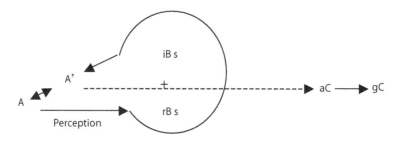

Figure a.1   ABC model (De Moor, 1995).

In this model an individual perceives information, represented by A, within an objective reality. This perception is filtered through a number of "lenses" (B) and transformed into a meaningfully constructed A' reality. This reality has consequences (C) at the affective level (aC) of feelings and emotions. Through those feelings and emotions, this reality also affects behavior (gC). Thus, the individual becomes, in De Moor's own words, a "spider in a web of self-spun realities." He compares this to the Thomas theorem, which runs "*If men define situations as real, they are real in their consequences.*"

In the process of converting information A to a constructed reality A', use is made of information about reality that has already been retained in the past. To illustrate this, De Moor uses the image of the "lenses." Lenses are hypothetical constructs. They cannot be measured or observed directly. The lenses people look through are composed of convictions, attitudes, norms, and values. De Moor distinguishes between environmental lenses, which contain information about the world around us, and "self-image" lenses, which contain information about ourselves. De Moor makes a further distinction between rational and irrational lenses. Rational lenses are realistic; irrational lenses are harmful to the people involved, cannot be validated and are interchangeable with other lenses. Irrational lenses cause people stress.

The perceived reality A' evokes feelings in people. These affects consequently give rise to behavior. That behavior elicits feedback from one's environment and this leads to an assessment of one's behavior in relation to the meaning A' and consequently to a reconstruction of A'.

Sensemaking According to Weick

The process that De Moor calls common sensemaking is called sensemaking by Karl Weick (1979). For him, sensemaking is a collective social process in which people create their own reality as they interact and act upon this reality.

Weick distinguishes three phases within the process of sensemaking: enactment, selection, and retention. Enactment is the phase in which information is actively identified and selected. Weick refers to this process as "bracketing" (adapted from Schultz, 1967) or a "punctuation of contexts." In the selection phase, the information that is bracketed is placed alongside diagrams/constructs that are already present in people's minds, a process that continues until an explanation is found. The explanation that is found is retained in the retention phase. This retention phase consequently acts as a guideline with regard to what is identified in the enactment phase and what serves as comparative material in the selection phase.

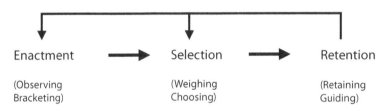

Figure a.2 Sensemaking (adapted from Weick, 1979).

In Weick's theory, this sequence applies to individuals who use it to create meaning for themselves and with each other. For Weick, the essence of sensemaking lies in the fact that it is a social process that people create together.

As in De Moor's theory, for Weick the process of sensemaking starts with the moment at which people notice something in their surroundings that is unfamiliar to them. He identifies three types of disturbances:

1. unexpected occurrences
2. occurrences that are expected but do not take place
3. processes that are interrupted.

In each of these cases, attention is drawn to something that is unfamiliar and to which meaning must be assigned. Not every disturbance is noted. Only that which is registered as a construct in the retention phase and which actually allows space for identifying the disturbance is noticed.

Sensemaking is an "ongoing process." The phases of enactment, selection, and retention do not play out in people's heads in a perfectly sequential manner, neatly arranged according to topic. In practice, all the phases take place simultaneously, they involve multiple occurrences at the same time, and they involve multiple people. It is not really possible to identify a beginning or an end within the stream of people's sensemaking. What can be identified is "interpunction" (Watzlawick, 1974): a moment within the sensemaking process at which people either have an occurrence start or choose a certain focus. At such moments, people "*chop out moments of continuous flows and extract cues from those moments*" (Beyer in Weick, 1995, p. 111).

For Weick, sensemaking forms the link between human action and cognition. Through interaction with their environment, people attribute meaning to occurrences that take place within it. For Weick, the most important component of interaction is "action." Cognition and action are connected to each other through sense-giving. Weick shows that people can only really know and adjust their meaning through their actions. To illustrate this, he uses the following sentence: "*How can I know what I think until I see what I say?*" (1979, p. 5) By acting (selection) we can learn to understand that what we see but do not (yet) understand (enactment) and retain it in our memory (retention). For this process we need the world around us. It is only by interacting with the world around us that we gather this information.

Types of Sensemaking

Weick makes a distinction between two types of sensemaking: "belief-driven processes" and "action-driven processes." Belief-driven processes are sense-giving processes that stem from the beliefs that people hold with regard to the world, as these beliefs are stored in people's minds in the form of ideologies, values, paradigms, and similar notions. These constructs and the beliefs that inform them guide the things people notice and the way in which they interpret events. As Weick (1995) states, "*To believe is to notice selectively*" (p. 133).

In belief-driven processes, people use two strategies for attributing meaning to their environment: arguing and "expecting." Arguing involves testing beliefs in interaction with others, resulting in rejection or reinforcement, with the aim of reducing contradictions and ambiguity. The second process, of "expecting," is aimed at confirming beliefs. That which is expected is filtered from the environment. People tend to focus more on confirming their beliefs than on questioning them. This type of sensemaking comes closest to the phenomenon of the self-fulfilling prophecy: one's expectations form the starting point for what one sees around you, how one interprets it, and how one reacts to it.

Action-driven processes stem from actions, with the objective of linking actions to presuppositions. Again, Weick makes a further distinction between two different strategies, that of "behavioral commitment" and manipulation. Behavioral commitment comes about when the person involved feels responsible for their actions (commitment). When someone is bound to that behavior, for example, because the behavior was clear to their environment and because it cannot be undone, they then need to find an explanation to rationalize the behavior. In that kind of situation, sensemaking is focused on finding an answer to the question why the behavior occurred: it accounts for displayed behavior. It is a way of reducing the cognitive dissonance (Festinger, 1964) that may have arisen through action.

Manipulation provides answers to the question of what happened. Where behavioral commitment provides answers to the question of why one's own behavior happened, manipulation provides answers to the question of what happened in the world outside. Manipulation involves selecting those particular facets from one's environment that make the world understandable. It is a process through which one creates one's own environment.

Organizational Culture

Seen from a social-constructivist point of view, an organization is a social reality that is continually being constructed or enacted by its members as they communicate and interact (De Moor, 1995). According to De Moor, collective sensemaking follows a process that is similar to the steps indicated in the individual ABC model. Through interaction, personal realities converge, become further attuned to each other, and a degree of overlap arises. This movement of convergence leads to a more or less shared reality. According to De Moor, organizational culture takes shape both on an abstract/mental level and on a concrete/observable level.

Figure a.3   Organizational culture (De Moor, 2005).

As a consequence, organizational culture is composed of an organizational ideology (the entirety of collective mental images about the organization that people have formed together) and an organizational climate (the behavioral patterns that have taken shape in the form of artifacts). The ideology is connected to the climate by way of collective behavioral choices and the process of organizing and sensemaking.

There is a difference in the way in which Weick and De Moor view the causal relation between organizational ideology and organizational climate. According to De Moor, the organizational ideology steers the organizational climate by way of collective behavioral choices: the construct precedes action. In Weick's view, on the contrary, action precedes the construct: meaning is created afterward. We think that in Weick's version of the above model, the arrows would point in the opposite direction.

Aspects of Sensemaking Processes and Constructs

According to Weick (1979), sensemaking processes are cognitive and social. They are cognitive because sensemaking processes create constructs based on information. The social aspect lies in the fact that meaning is created in a social context and through interaction between people. Homan attributes a cognitive, a behavioral and physical aspect to sensemaking processes. For him, the cognitive aspect consists of mental models, and the behavioral aspect consists of the behavioral patterns linked to cognition. The physical aspect consists of the physical processes (e.g., hormonal and neurological processes) that precede the cognitive and behavioral processes. Hosking and Morley (1991) attribute a cognitive and a political aspect to sensemaking processes. They argue that people take power into account in sensemaking processes and constructs.

Constructs

Sensemaking produces constructs. De Moor (1995) argues that within organizational culture, constructs can be characterized according to ideology and climate. In his view, individual constructs are incorporated in lenses (see preceding "Individual Sensemaking" section).

Weick (1995) divides constructs into ideologies, third-order controls, paradigms, theories of action, traditions, and stories. These all form frameworks that guide sensemaking processes.

Ideologies

Weick (1993) uses Beyer's definition of ideologies as *"shared, relatively coherent interrelated sets of emotionally charged beliefs, values and norms that bind some people together and help them make sense of the world"* (p. 234). Ideologies are a way for organizations to use values to connect realities and relationships.

Third-Order Control
Third-order control is an organizational control mechanism that consists of assumptions that are taken for granted. First-order control consists of direct supervision, second-order control of the rules and procedures that have been developed, and third-order controls are "taken for granted assumptions" in the organization (Schein, 2006). Third-order control is also referred to as the "professional blind spot." Weick connects it to the assumptions that are part and parcel of organizational cultures. These three forms of "control" are mechanisms that are used in organizations in order to create order in work practices. Third-order control is especially important in relation to non-routine work practices.

Paradigms
Paradigms consist of broad intellectual visions—worldviews—into which fundamental cognitive schemas and values are closely woven. Paradigms are the convictions that provide direction for professional behavior.

Theories of Action (Argyris)
Theories of action are sets of rules that individuals use to shape their own behavior and to interpret the behavior of others. They are cognitive organizational structures that direct the behavior of people in organizations. Something that adds to the complexity of these theories of action is the fact that the things people say diverge from what they actually do. That is why Argyris makes a distinction between espoused theories and theories in use. Theories are resistant to change.

Traditions
A tradition is a "belief" from the past that is inherited by future generations. Traditions can consist of all kinds of images.

Stories
"Most organizational realities are based on narration" (Weick, 1995, p. 127). Stories are "guides of conduct" for people in organizations.

In short, constructs, the results of sensemaking processes, can be seen in organizations in many different forms. The inventory drawn up by Weick does not constitute a coherent whole, but should rather be seen as an outline of different concepts taken from the literature that can be placed in the category of "constructs."

Organizational Landscape
Homan (2006) describes the way in which an organizational landscape comes into being through sense-giving processes. In sensemaking processes, "patches" of meaning are formed in local groups through the convergence of meaning. These patches consist of converged meaning of people in an organization.

Once they have formed, these patches crystallize, because they are continually confirmed in interaction. In the crystallization process, meaning also becomes a part of the identity of local groups. Patches become communities, and in the crystallization process it becomes increasingly clear who belongs to which community. Moreover, communities have their own regime protectors who appoint themselves as the protectors of the meaning that has been created.

Organizations contain multiple patches of meaning. These patches can converge and diverge, and they can be connected to a greater or lesser degree. The connection between the different communities is characterized by two factors: a K factor and an R factor. The K factor indicates the degree of connection between two communities: the higher K is, the more links there are between them and the closer they are. The R factor is the regulatory factor and this gives an indication of the power relations between communities. Homan calls the web that is woven between communities the "social fabric."

If there is an ongoing convergence of meaning, this produces tight couplings. If meanings diverge, this results in loose couplings. This degree of connection also applies to communities: when communities are relatively independent of each other, one can speak of "loose couplings" (Weick). When communities are dependent on each other, resulting in close relations, we can speak of tight couplings. These mechanisms produce an organizational landscape in which different communities, with their own constructs, are more or less closely connected to each other.

According to Homan (2006), a collective construct is also formed alongside the meaning that is created in the local communities. On an organizational level, this collective construct consists of converged meaning. According to Homan, organizations' collective constructs are flexible and adapt to important changes. Through interaction, the collective construct and the local communities exert a mutual influence on each other.

Appendix B: Overview

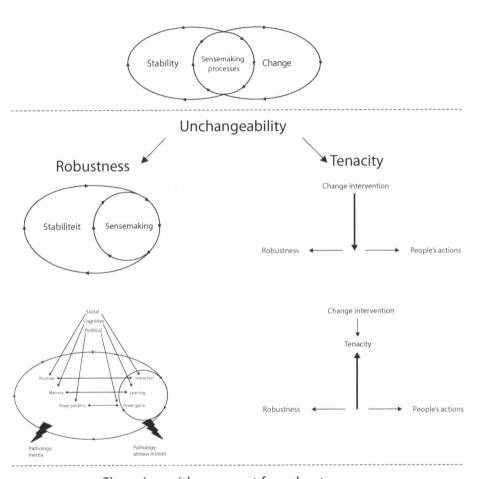

Changing with a respect for robustness

	Focus	Changeability of sensemaking processes	Robust side of sensemaking processes
Method			
Facilitating		Creating preconditions for sensemaking 1	Offering new meaning: sensegiving 3
Coercive		Organizing reflection 2	Enforcing new meaning 4

About the Authors

Jaap van 't Hek

Jaap van 't Hek (1953) studied human resources management in Amsterdam in the 1970s. At the end of 1980s, he did a two-year degree in Organizational Studies at the SIOO institute. Among other things, he has worked as a policy official for the Municipality of Amsterdam and as an organizational consultant for several consulting firms. In 1997 he shifted the focus of his activities to interim management by starting Organisatievragen, in which Leike van Oss joined him a year later. He has worked as an interim director for a range of organizations including a number of municipalities, the Royal Netherlands Academy of Arts and Sciences, Pax Christi, the Police Academy, the Ministry of Education, and the Amsterdam School of the Arts. Since 1992, he has been a lecturer at the SIOO institute. Jaap has a penchant for complex organizations and enjoys working with professionals ranging from gardeners to university professors, who are no big fans of "the management" but need to work with them nonetheless.

Leike van Oss

Leike van Oss (1964) studied natural medicine and social psychology. For a number of years she ran a practice in homeopathy and acupuncture and worked at GITP International as an organizational development consultant. Since 1998, she has formed Organisatievragen together with Jaap van 't Hek, and has worked as an organizational consultant, interim manager and coach. Leike works mainly with knowledge-intensive organizations and with national and local government organizations. Leike deals with assignments that revolve around loosening up persistent patterns, and which are often related to administration and staff-line relations within organizations. Leike is fascinated by the question of why it is that the way organizations actually function diverges so much from the declared management objectives, and why change processes often proceed so differently than hoped for or anticipated.

She teaches and gives guest lectures at a number of institutions, including SIOO, VU University Amsterdam, Radboud University Nijmegen, and Schouten & Nelissen University.

For more information about the authors, please visit www.organisatievragen.nl and www.onveranderbaar.nl.

About the Contributing Authors

Hans Bennis

Hans Bennis studied Dutch and general linguistics in Amsterdam, at the University of Amsterdam, and VU University. After a period spent working as a researcher at the University of Amsterdam, he was appointed as associate professor in the Dutch department of Leiden University. His specialization was in the field of generative grammar, with a particular focus on the grammar of small words. In 1986, he obtained a doctorate in Philosophy at Tilburg University with a dissertation entitled "Gaps and Dummies." He stayed at Leiden University until 1998. In later years, mainly in his capacity of director of the inter-university research institute, the Holland Institute of Generative Linguistics, he continued combining science and management in his new position as the director of the Meertens Institute, part of the Royal Netherlands Academy of Arts and Sciences (KNAW). In 2000, he was appointed as professor in Language Variation at the University of Amsterdam. Hans is a member of many national and international committees in the field of language and culture, but his foremost passion is linguistic research. His CV and list of publications can be found on the Meertens Institute website: www.meertens.knaw.nl.

Frank Bijdendijk

After graduating from the Montessori Lyceum in Rotterdam, Frank Bijdendijk (1944) studied physics and later architecture at what was then known as the Polytechnical Hogeschool in Delft. He went on to spend six years as a project manager with the construction company Verenigde Bedrijven Bredero in Utrecht (projects: Hoog Catherijne and Scheveningen), one year as the head of project coordination at MAB Projectontwikkeling in The Hague and four years as a technical director at Lunetten BV in Utrecht, which was tasked with developing the Lunetten residential district for the Utrecht housing corporations. Since September 1, 1982, he has been the general director of the housing association Het Oosten in Amsterdam, and, after a merger with the Algemene

Woningbouw Vereniging on July 1, 2008, of the newly formed association Stadgenoot. In January 2007 he received the Award for Corporate Director 2006, awarded by Aedes Directeurencontact. On September 1, 2007, Frank Bijdendijk was appointed a Knight in the Order of Oranje Nassau, for services relating to public housing and his personal efforts dedicated to social causes.

Piet Boonekamp

Piet Boonekamp (1950) studied biochemistry at Leiden University in the 1970s, then fulfilled his national service requirements by studying cerebrospinal meningitis at the National Institute for Public Health and the Environment in Bilthoven and went on to carry out PhD research into bone metabolism and immunology at Leiden University and the University of Missouri in Kansas City. In the mid-1980s, his knowledge of immunology brought him to Wageningen for research into plant disease diagnostics, and he continued in this line of research as the head of the plant disease department of the Flower Bulb Research Center at Lisse. At the end of the 1990s he returned to Wageningen to act as a change manager in a merger of the three existing plant research institutes, and was then appointed as the manager of the Business Unit "Biointeractions," which was created as a result of the merger and which conducts research into subjects including the interaction between plants and pathogens. What Piet likes best is a combination of scientific content, stimulating researchers, managing a large group of individually-minded professionals, and setting up large-scale international research programs that create impact for the stakeholders involved.

Christien Brinkgreve

Christien Brinkgreve (1949) studied sociology at the University of Amsterdam, was a professor of Women's Studies at the University of Nijmegen for a number of years, and a professor of Social Sciences at the University of Utrecht from 1990 onward. She has published many articles on the relations between men and women in academic and literary journals, has written a large number of books, and wrote a column for one of the Netherlands' major daily newspapers for ten years. She now works partly as a freelancer, is frequently engaged as a public speaker, and is interested in the connections between social developments and people's intimate personal lives, their self-awareness, and emotions.

Saskia van Dockum

Saskia van Dockum (1965) studied archaeology at Leiden University, specializing in provincial Roman archaeology in the Netherlands. Previously she has held the post of provincial archaeologist of Utrecht, and various positions with the Netherlands' former State Service for Archaeological Investigations, including that of head of Preservation and director. In 2005, she moved to a different heritage sector, that is, the archives and records sector. As the director of the Utrecht Archives, she is responsible for opening an innovative, accessible public center in the Utrecht Museum Quarter. This makes the Utrecht Archives a new or renewed player in Utrecht's cultural playing field and demands a changeover on the part of her employees. Saskia enjoys kindling enthusiasm in her organization's activities and giving her employees full responsibility regarding their own contribution to the organization as a whole.

René Gude

René Gude (1957) studied social geography for a number of years before he discovered his true field: philosophy. Within the field, epistemology became his greatest love and the philosophers Descartes and Kant came to be his faithful companions. This situation has remained unchanged since. After graduating, he started working for the Netherlands' most widely distributed magazine on philosophy, *Philosophy Magazine*, first as an editor and later as the editor in chief/publisher. This was followed by a period during which he worked as a strategic consultant for L. J. Veen Publishers and as a freelance columnist and publicist. Since 2002, he has been the director of the International School of Philosophy in Leusden. There, philosophy enthusiasts have been able to follow weekend courses for nearly a century now, and, starting a decade ago, businesses have been trained in the constants of organizational philosophy.

Jan van Hooff

Jan van Hooff (1936) grew up in the family that owned Burger's Zoo in Arnhem. Therefore it was no surprise that he went on to study biology, which he did in Utrecht and Oxford (under Niko Tinbergen). From 1980 to 2001 he was a professor of Behavioral Biology at the University of Utrecht. His special interests are social behavior and the social organization of primates in particular. He obtained his doctorate with a comparative study of facial expressions as a means of social communication, and more specifically of the evolution of laughter and smiling. His later research and that of his department focused on the regulation of social relations (including conflict regulation and social harmony in animal communities), studies in captive colonies (the famous chimpanzee colony at Burgers Zoo in Arnhem, for example) as well as on the socio-ecology of primate species in field studies (including studies of orangutans in Indonesia). Jan van Hooff is a member of the Royal Netherlands Academy of Arts and Sciences (KNAW).

Theo Mulder

Theo Mulder is a neuropsychologist. He studied psychology at Nijmegen University and was awarded his doctorate for a dissertation titled "The Re-Learning of Motor Control Following Stroke: Clinical and Experimental Studies." In 1994, he was appointed professor of Rehabilitation Research, and was connected to the Neurological Institute of the University of Nijmegen. From 1987 to 1999 he was also attached to the Sint Maartenskliniek (SMK) in Nijmegen, where he founded SMK-Research, a center for the study of motor disorders. In 1999, he left Nijmegen for Groningen to take up the position of professor of Human Movement Sciences and of director of the Center for Human Movement Studies. His primary area of interest is the flexibility of neural motor control (neural plasticity) related to questions regarding learning and recovery after damage to the nervous system. After a period of seven years in Groningen, he accepted the position of Director of Research at the Royal Netherlands Academy of Arts and Sciences (KNAW), where he is responsible for the nineteen Royal Academy Institutes in the Netherlands (1300 fte). He is still attached to the University of Groningen on the basis of a zero-time appointment. Furthermore, Theo Mulder is fascinated by the exciting relation between art and science.

Louise Vet

Louise Vet (1954) studied biology and conducted her PhD at Leiden University. In 1984, she joined the faculty of Wageningen University, and was appointed professor of Evolutionary Ecology in 1997. She is well known internationally for her work on parasitoid behavior and plant-parasitoid interactions. In 1996 the International Society of Chemical Ecology awarded her the Silverstein-Simeone lecture award. Since 1999 she has been the director of the Netherlands Institute of Ecology (NIOO-KNAW). The NIOO conducts research in the field of marine, freshwater, and terrestrial ecology. She carries out her own research into multitrophic interactions at the NIOO and at Wageningen University, where she maintains her professorate in Evolutionary Ecology. Apart from the University of Leiden and Wageningen University she conducted research at the University of Riverside California, Simon Fraser University Vancouver, and University of Toronto, Canada. In 2004 she was appointed a member of the Royal Netherlands Academy of Arts and Sciences (KNAW), and in 2005 she and two of her colleagues at Wageningen University were the first Dutch scientists to be awarded the UK Rank Prize for Nutrition. She has published over 180 chapters in books and articles in international peer-reviewed journals. She serves on a variety of national and international boards and committees. Louise Vet sees it as her mission to communicate the importance of biology and more specifically ecology to the broader public via the media, through columns, lectures (e.g., TEDxAmsterdam) and through the popular televised science quiz "Hoe?Zo!"

References

Ardon, A. J. (2006). Leiderschap en interventies in stagnerende veranderprocessen. *M&O, 60*(6), 5–23.

Argyris, C. (1991). Teaching smart people how to learn. *Harvard Business Review, 69*(3), 99–109.

Argyris, C. (1999). *On organizational learning* (2nd ed.). Oxford: Blackwell.

Bahlmann, J. P., Meesters, B. A. C., & Nunnink, H. E. W. (1998). *De organisatie die nooit bestond; een zoektocht naar vormen van organisatie.* Schoonhoven: Academic Services.

Bakema, F. (2006). *The emergence of a competitive group competence in a research group.* Eindhoven: Ecis.

Bateson, G. (1984). *Het verbindend patroon.* Amsterdam: Bert Bakker.

Becker, H. S. (1988). *Tricks of the trade.* Chicago: University of Chicago Press.

Bijdendijk, F. (2006). *Met andere ogen.* Amsterdam: Het Oosten.

Boonstra, J. J. (2000). *Lopen over water, over dynamiek van organiseren, vernieuwen en leren.* Amsterdam: Vossiuspers AUP.

Boonstra, J. J., Steensma, H. O., & Demenint, M. I. (1996). *Ontwerpen en ontwikkelen van organisaties; theorie en praktijk van complexe veranderingsprocessen.* Maarssen: Elsevier/de tijdstroom.

Braudel, F. (1993). *A history of civilizations.* New York: Penguin.

Brown, R. (2000), *Group processes* (2nd ed.). Oxford: Blackwell.

Caluwé, L. de, & Vermaak, H. (2006). *Leren veranderen, een handboek voor de veranderkundige.* Deventer: Kluwer.

Capra, F. (1996). *The web of life.* New York: Random House.

Clegg, S. R. (1989). *Frameworks of power.* Thousand Oaks, CA: Sage.

Clegg, S. R. (1990). *Modern organizations, organization studies in the postmodern world.* Thousand Oaks, CA: Sage.

Coutu, D. (2002). Edgar H. Schein: The anxiety of learning. *Harvard Business Review, 3,* 100–108.

Diamond, J. (1998). *Guns, germs and steel: A short history of everybody for the last 13,000 years.* London: Vintage.

Dongen, H. J. van, de Laat, W.A.M., & Maas, A. J. J. A. (1996). *Een kwestie van verschil.* Delft: Eburon.

Editorial. (2007, September 27). Kabinet roelbekkert. *NRC Handelsblad,* 07.

Gioia, D. A., & Chittipeddi, K. (1991). Sensemaking and sensegiving in strategic change initiation. *Strategic Management Journal, 12,* 443–448.

Goffman, E. (1959). *The presentations of self in everyday life.* New York: Anchor Books.

Gray, J. (2002). *Straw dogs: Thoughts on humans and other animals.* London: Granta.

Harrison, R. (1970), Choosing the depth of organizational intervention. *Journal of Applied Behavioral Science, 6*(2), 189–202.

Hatch, M. J. (1997). *Organization theory, modern symbolic and postmodern perspectives.* Oxford: Oxford University Press.

Heijden, C. van der. (2007). *Het zand in de machine, managerscultuur in Nederland.* Amsterdam/Antwerpen: Contact.

Hermans, J. M., & Hermans-Jansen, E. (1995). *Self-narratives: The construction of meaning in psychotherapy.* New York: Guildford Press.

Homan, T. (2006). *Organisatiedynamica, theorie en praktijk van organisatieverandering.* Den Haag: Academic Service.

Hosking, D. M., & Morley, I. E. (1991). *A social psychology of organizing: People, processes and contexts.* New York: Harvester Wheatsheaf.

Kahn R. L., & Boulding, E. (1966). Power and conflict in organizations. The Economic Journal, *76*(301), 110–112

Kampen, J., & Schuiling, G. J. (2005). Verwaarloosde organisaties: (her)opvoeden, een vergeten taak van de manager. *M&O, 59*(5), 30–50.

Lam, A. (2004). Knowledge, learning in organizational embeddedness: A critical reflection. In J. Boonstra (Ed.), *Dynamics of organizational change and learning* (pp. 177–196). Chichester: John Wiley.

Luhmann, N. (1984). *Social systems.* Stanford, CA: Stanford University Press.

Maturana, H. R., & Varela, F. J. (1980). *Autopoiesis and cognition: The realization of the living.* Dordrecht: D. Reidel Publishing.

Metze, M. (2009–2010). *Veranderend getij: Rijkswaterstaat in crisis.* Amsterdam: Balans.

Mintzberg, H. (2004). *Managers not MBAs: A hard look at the soft practice of managing and management development.* San Francisco: Berret-Koehler.

Moor, W. de (1998). Organisatieverandering, een constructionistisch perspectief. *M&O, 52*(6), 45–61.

Moor, W. de (2005). *Het proces van organiseren, individueel en sociaal-constructionisme, praktijkmodellen.* Apeldoorn: Garant.

Morgan, G. (1997). *Images of organization.* Thousand Oaks, CA: Sage.

Mulder, M. (1972). *Het spel om macht; over verkleining en vergroting van machtsongelijkheid.* Meppel: Boom.

Mulder, M. (1977). Omgaan met macht, ons gedrag met elkaar tegen elkaar. Amsterdam-Brussel: Elsevier

Musil, R. (1996). *The man without qualities* (B. Pike & S. Wilkins, Trans.). London: Pan Macmillan. (Original work published 1930)

Nonaka, I., & Takeuchi, H. (1995). De kenniscreërende onderneming; hoe Japanse bedrijven innovatieprocessen in gang zetten. Schiedam: Scriptum Management.

Oss, van L. (2002). De bril van de ander. In J. van den Oever & M. Otto (Eds.) *Organisatiediagnostiek: betekenis geven aan gedrag in organisaties.* Deventer: Kluwer.

Otto, M. M. (2000). *Strategisch veranderen in politiek bestuurde organisaties.* Assen: Van Gorcum.

Otto, M. M., & Leeuw, A. C. de (1994). *Kijken, denken, doen. Organisatieverandering: manoeuvreren met weerbarstigheid.* Assen: Van Gorcum.

Pasmore, W. A., & Fagans, M. R. (1992). Participation, individual development and organizational change: a review and synthesis. *Journal of management, 18*(2), 375–397.

Peter, L. J., & Hull, R. (1969). *The Peter principle.* New York: W. Morrow.

Poole, M. S., Ven, A. H. van de, Dooley, K., & Holmes, M. E. (2000). *Organizational change and innovation processes: Theory and methods for research.* Oxford: Oxford University Press.

Ruijters, M. (2006). *Liefde voor leren; over diversiteit van leren en ontwikkelen in organisaties.* Deventer: Kluwer.

Schein, E. H. (1992). *Organizational culture and leadership* (2nd ed.). San Francisco: Jossey-Bass.

Schein, E. H. (1999). *Corporate culture survival guide: Sense and nonsense about culture change.* San Francisco: Jossey-Bass.

Schön, D. A. (1983). *The reflective practitioner: How professionals think in action.* New York: Basic Books.

Schreuder, A. (2008, January 19). Je moet leren meebewegen. *NRC Handelsblad,* Z04.

Scott-Morgan, P. (1995). *De ongeschreven regels van het spel.* Groningen: BoekWerk.

Senge, P. (2006). *The fifth discipline: The art and practice of the learning organization* (Rev. ed.). New York: Random House.

Skrine, P. (1987). An age of exuberance, drama and disenchantment. *The UNESCO Courier,* March, 4–8.

Stacey, R. D. (1992). *Managing the unknowable, strategic boundaries between order and chaos in organizations.* San Francisco: Jossey-Bass.

Ten Have, S. (2002). *Key management models: the management tools and practices that will improve your business.* Harlow: Pearson Education.

Termeer, C. J. A. M. (2001), Hoe je een boer aan het lachen krijgt. In Abma T. & R. in 't Veld (Eds), *Handboek Beleidswetenschap.* Amsterdam: Uitgeverij Boom, 366–375.

Tosey, P. (2006). *Bateson's levels of learning: A framework for transformative learning?* Paper presented at universities forum for Human Resource Development conference. Universiteit van Tilburg. Retrieved February 4, 2011, from http://epubs.surrey.ac.uk/1198/

Valens, P. (1989). Brown Jackets and blue jackets. Retrieved February 4, 2011, from http://www.valens.nl/enhtml/105.html.

Voigt, R., & Spijker, W. van (2003). *Spelen met betekenis: verhalen over succesvol vernieuwen bij de overheid.* Assen: Koninklijke van Gorcum.

Waal, F. de (1998). *Chimpanzee politics, power and sex among apes* (Rev. ed.). Baltimore, MD: Johns Hopkins University Press.

Waal, F. de (2005). *Our inner ape.* New York: Riverhead Books.

Watzlawick, P. (1967). *Pragmatics of human communications.* New York: W.W. Norton & company.

Watzlawick, P. (1990). *Münchhausens's pigtail or psychotherapy an "reality."* New York: W.W. Norton & Company.

Watzlawick P., Weakland J. H., & R. Fish (1974). *Change: Principles of problem formation and problem resolution.* New York: W.W. Norton & Company.

Weick, K. E. (1977). Enactment processes in organizations. In B. M. Staw & G. Salancik (Eds), *New directions in organizational behaviour* (pp. 267–300). Chicago: St. Clair.

Weick, K. E. (1979). *The social psychology of organizing* (2nd ed.). New York: McGraw-Hill Inc.

Weick, K. E. (1988). Enacted sensemaking in crisis situations. *Journal of Management Studies, 25*(4), 305–317.

Weick, K. E. (1990). The vulnerable system: an analysis of the Tenerife air disaster. *Journal of Management, 16*(3), 571–593.

Weick, K. E. (1993a). Collective mind in organizations: Heedful interrelating on flight decks. *Administrative Science Quarterly, 38,* 357–381.

Weick, K. E. (1993b). The collapse of sensemaking in organizations: the Mann Gulch Disaster. *Administrative Science Quarterly, 38,* 628–652.

Weick, K. E. (1995). *Sensemaking in organizations.* Thousand Oaks, CA: Sage.

Weick, K. E. (2001). *Making sense of the organization.* Malden, MA: Blackwell Publishers.

Weick, K. E., & Quinn, R. (2004). Organizational change and development, episodic and continuous changing. In J. J. Boonstra (Ed.), *Dynamics of organizational change and learning* (pp. 177–196). Chichester: John Wiley & Sons Ltd.

Weick. K. E., & Sutcliffe, K. M. (2001). *Managing the unexpected.* San Francisco: Jossey-Bass.

Witman, Y. (2008) *De medicus maatgevend, over leiderschap en habitus.* Assen: Koninklijke van Gorcum.

Index